# MALONDIALDEHYDE (MDA)

## STRUCTURE, BIOCHEMISTRY AND ROLE IN DISEASE

# BIOCHEMISTRY RESEARCH TRENDS

Additional books in this series can be found on Nova's website
under the Series tab.


Additional e-books in this series can be found on Nova's website
under the e-book tab.

# MALONDIALDEHYDE (MDA)

## STRUCTURE, BIOCHEMISTRY AND ROLE IN DISEASE

JACKSON CAMPBELL
EDITOR

*New York*

### NOTICE TO THE READER



**Library of Congress Cataloging-in-Publication Data**

ISBN: 978-1-63482-793-5
Library of Congress Control Number: 2015938204

*Published by Nova Science Publishers, Inc. † New York*

# CONTENTS

# **PREFACE**

Malondialdehyde (MDA) is one of the better-known secondary products of lipid peroxidation, and it can be used in biomaterials as an indicator of cell membrane injury. The first chapter offers an overview of the implications of MDA on oxidative stress, and therefore in neurogenerative diseases, as well as collects information of several natural products that exert beneficial properties on oxidative stress-related diseases by modulating MDA levels. The second chapter in this book focuses on the role of malondialdehyde in the development of insulin resistance in obesity. The chapter also comprehensively summarizes the pathophysiological roles of malondialdehyde and its impact on cognitive function in obesity. The third and final chapter looks at MDA as a marker of lipid peroxidation, in the internal organs of of Wistar rats. MDA is the final product in the lipid peroxidation process, and the degree of lipid peroxidation, oxidative stress, can be evaluated by measuring the level of MDA in different tissues. The potential protective effects of O- and S- donor supplements (glutathione and lipoic acid) on lipid peroxidation effects were described as well.

Chapter 1 – Oxidative stress results from pro-oxidant/antioxidant imbalance leading to an overproduction of reactive oxygen species (ROS) which can cause oxidative damage to cell structures, including alterations in membrane lipids and proteins, and even trigger to cell death. Oxidative stress contributes to the progressive neurodegeneration associated to diseases such as Alzheimer´s disease and Parkinson´s disease, among others. Within the harmful effects at cellular level of oxidative stress, lipid peroxidation is a free radical oxidation (chain reaction) of polyunsaturated fatty acids such as linoleic acid or arachidonic acid, and it has been related with various pathologies and disease status mainly because of the oxidation products formed during the process. These products include reactive aldehydes such as malondialdehyde and 4-hydroxynonenal (HNE). Malondialdehyde (MDA) is one of the better-known secondary products of lipid peroxidation, and it can be used in biomaterials as an indicator of cell membrane injury. Due to its reactivity, MDA adducts can react with proteins and DNA, promoting intramolecular or intermolecular protein/DNA crosslinking that may induce profound alteration in the biochemical properties of biomolecules, and then facilitate the development of various pathological state. The search of exogenous antioxidants that counteract oxidative stress damage, thus reducing lipid peroxidation and MDA formation by different mechanisms, emerges as a promising tool for the finding of neuroprotective agents. Present chapter offers an overview of the implications of MDA on oxidative stress, and therefore in neurodegenerative diseases, as well as collects information of several natural products that exert beneficial properties on oxidative stress-related diseases by modulating MDA levels.

Chapter 2 – The incidence and level of obesity has increased dramatically in many countries around the world. Obesity escalates the risk of increased oxidative stress, which is believed to be one of the major underlying causes of many pathological conditions including insulin resistance, type II diabetes mellitus and cognitive decline. Malondialdehyde acts as a biomarker for oxidative stress. There is a great deal of evidence to show the impact of the involvement of malondialdehyde in many pathological conditions. This chapter focuses on the role of malondialdehyde in the development of insulin resistance in obesity. Recent research in obese models has shown that increased oxidative stress can also lead to brain insulin resistance in addition to peripheral insulin resistance. The impairment of brain insulin sensitivity has been strongly implicated in the development of cognitive decline. This chapter also comprehensively summarizes the pathophysiological roles of malondialdehyde and its impact on cognitive function in obesity.

Chapter 3 – In this study was determined a content of MDA, as a marker of lipid peroxidation, in the internal organs (liver, kidneys, brain and pancreas) of heavy metal intoxicated Wistar rats. The potential protective effects of O- and S- donor supplements (glutathione and lipoic acid) on lipid peroxidation effects were also examined. The content of malondialdehyde (MDA) was increased several times in the investigated tissues of exposed rats. The toxic effects of investigated metals being more pronounced in liver, respectively by metal: Pb > Cd > Cu. In other organs the toxic effect of metals, according of the level of MDA is respectively: Cd > Pb > Cu. The treatment of intoxicated animals with glutathione and lipoic acid drastically suppressed lipid peroxidation. These supplements can form via their O- and S- donor atoms stable associations with the ions of these metals, thus blocking the heavy metals and reducing their toxic effects in the following order Cd > Pb > Cu. FTIR analysis on the example of associates between copper and oxide and reduce form of lipoic acid, shows that reduced form react with S- donor atoms of free HS- groups, therefore oxide form react via O-donor atoms of carboxylic gropu. The intake of food rich with these supplements and of products of a similar structure may have a preventive effect (inhibition of lipid peroxidation), and significantly reduce the toxic influence of these metals.

In: Malondialdehyde (MDA)
Editor: Jackson Campbell

ISBN: 978-1-63482-793-5
© 2015 Nova Science Publishers, Inc.

*Chapter 1*

# INVOLVEMENT OF MALONDIALDEHYDE IN OXIDATIVE STRESS AND NEURODEGENERATIVE DISEASES

*C. Fernández-Moriano*, E. González-Burgos**
*and M. P. Gómez-Serranillos[†]*
Department of Pharmacology, Faculty of Pharmacy,
University Complutense of Madrid, Spain

## ABSTRACT

Oxidative stress results from pro-oxidant/antioxidant imbalance leading to an overproduction of reactive oxygen species (ROS) which can cause oxidative damage to cell structures, including alterations in membrane lipids and proteins, and even trigger to cell death. Oxidative stress contributes to the progressive neurodegeneration associated to diseases such as Alzheimer´s disease and Parkinson´s disease, among others.

Within the harmful effects at cellular level of oxidative stress, lipid peroxidation is a free radical oxidation (chain reaction) of polyunsaturated fatty acids such as linoleic acid or arachidonic acid, and it has been related with various pathologies and disease status mainly because of the oxidation products formed during the process. These

---

[*] Both authors contributed equally.
[†] Corresponding author.

products include reactive aldehydes such as malondialdehyde and 4-hydroxynonenal (HNE).

Malondialdehyde (MDA) is one of the better-known secondary products of lipid peroxidation, and it can be used in biomaterials as an indicator of cell membrane injury. Due to its reactivity, MDA adducts can react with proteins and DNA, promoting intramolecular or intermolecular protein/DNA crosslinking that may induce profound alteration in the biochemical properties of biomolecules, and then facilitate the development of various pathological state.

The search of exogenous antioxidants that counteract oxidative stress damage, thus reducing lipid peroxidation and MDA formation by different mechanisms, emerges as a promising tool for the finding of neuroprotective agents.

Present chapter offers an overview of the implications of MDA on oxidative stress, and therefore in neurodegenerative diseases, as well as collects information of several natural products that exert beneficial properties on oxidative stress-related diseases by modulating MDA levels.

# 1. OXIDATIVE STRESS, LIPID PEROXIDATION AND MALONYLDIALDEHYDE (MDA)

## 1.1. Oxidative Stress

Oxidative stress occurs when reactive oxygen species (ROS) (i.e., hydrogen peroxide, superoxide anion, hydroxyl radical and hypochlorite ion), which are by-products of normal cellular metabolic activities, surpass the enzymatic and the non-enzymatic antioxidant defenses (i.e., superoxide dismutase enzyme, catalase enzyme, glutathione peroxidase enzyme and glutathione) in their ability to neutralize them. As a consequence of this pro-oxidant/antioxidant imbalance, prolonged oxidative damage to cellular structures affects lipids of cell membranes, proteins and nucleic acids (DNA, RNA), and even triggers to cell death. Consistent experimental and clinical evidences indicate a prominent role of oxidative stress in the pathogenesis of a wide variety of diseases, including neurodegenerative disorders such as Alzheimer´s disease (AD), Parkinson´s disease (PD) and amyotrophic lateral sclerosis (ALS). Indeed, brain presents particular features which make neural tissues of this organ especially vulnerable to oxidative damage because of high content of polyunsaturated lipids, high oxygen consumption, low levels of antioxidant defenses, high intracellular concentrations of transition metals and reduced capacity to regenerate the neural tissue in comparison with other types of tissues [1-3].

As previously mentioned, lipids are major key targets for ROS damage and so, this chapter will be focused on these macromolecules, particularly malonyldialdehyde (MDA), and its role in oxidative stress and neurodegenerative diseases as well as information of several natural products that exert beneficial properties on oxidative stress-related diseases by modulating MDA levels.


## 1.2. Lipid Peroxidation

Lipid peroxidation is a complex process that refers to the oxidative degradation of lipids including polyunsaturated fatty acids (PUFAs), glycolipids, phospholipids and cholesterol, leading to the disruption of cellular membrane structure and function. There have been described three different mechanisms involving the oxidative deterioration of lipids: autoxidation (which is a free radical mediated process), photo-oxidation and enzymatic lipid oxidation [4-6].

### *Autoxidation: Non-enzymatic Lipid Peroxidation*

Autoxidation is the most common mechanism of lipid peroxidation; it is a free radical chain sequence process that occurs when atmospheric oxygen reacts spontaneously with lipids. Autoxidation consists of three different reaction steps: initiation, propagation and termination (Figure 1).

In the initiation step, free radicals, specially hydroxyl radical (HO$^\bullet$) and hydroperoxyl radical (HO$^\bullet_2$), abstract a hydrogen atom from a methylene carbon of unsaturated lipids leading to the formation of a lipid radical (L$^\bullet$). Subsequently, this lipid radical is stabilized by a molecular rearrangement to form a conjugated diene which reacts with molecular oxygen to give a lipid peroxyl radical (LOO$^\bullet$). In the propagation step, this lipid peroxyl radical (LOO$^\bullet$) removes an allylic hydrogen from a neighbouring lipid molecule producing a lipid hydroperoxide (LOOH) and new lipid radicals (L$^\bullet$), so self-propagating chain-reaction continues. The chain length in this propagation step depends on the fatty acid composition, oxygen levels, lipid/protein content ratio and the presence or absence of endogenous and exogenous antioxidants. In the termination step, lipid peroxidation ends when two peroxyl radicals (LOO$^\bullet$), two lipid radicals (L$^\bullet$) or one peroxyl radical (LOO$^\bullet$) and one lipid radical (L$^\bullet$) react to yield non radical products (stable end products).

Moreover, this reaction may also be interrupted in presence of antioxidants compounds such as α-tocopherol, which breaks lipid peroxidation reaction chain by donating hydrogen atoms to lipid peroxyl radical (LOO•) ensuing in the generation of nonradical products [4-6].

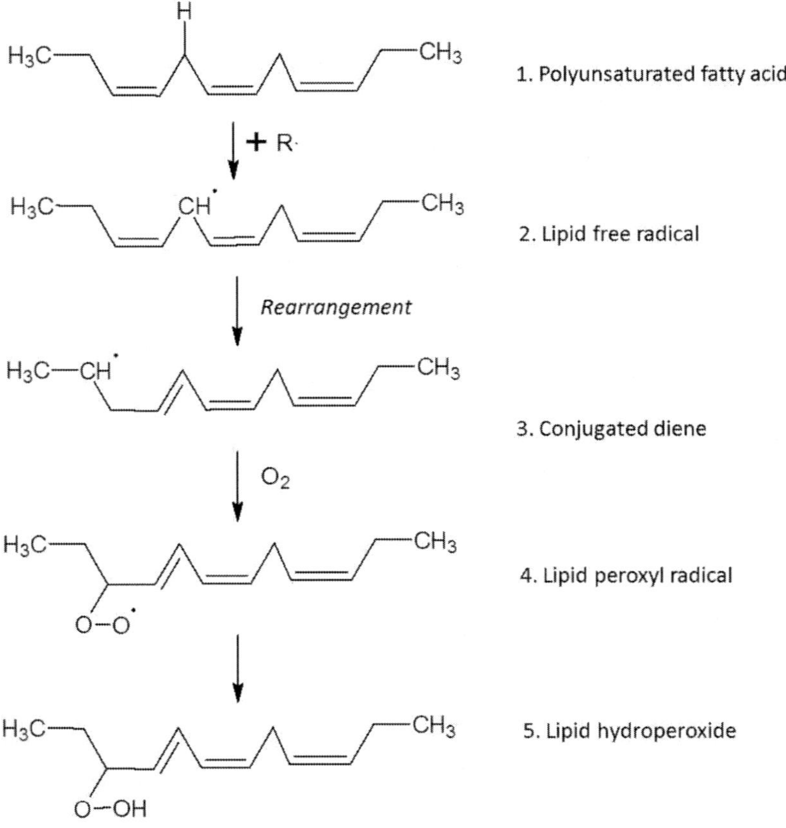


Figure 1. Autoxidation process of polyunsaturated fatty acids.

## *Photo-oxidation*

The singlet molecule oxygen ($^1O_2$), which is an excited stated of the dioxygen molecule ($O_2$) and is produced by photochemical sensitizers such as bilirubin, riboflavin and myoglobin, may react by addition with double bonds of lipids molecules and generate hydroperoxides (Figure 2). The photo-oxidation process is quicker than autoxidation process [6, 7]. As an example, Frankel et al., (1979) analyzed the autoxidized *versus* photosensitized oxidation of hydroperoxide isomers in unsaturated fats and esters by gas chromatography-mass spectrometry and they determined that the photo-oxidation of polyenes may be from 1,000 to 1,500 times quicker than its autoxidation reaction and the photo-oxidation of oleic acid may be 30000 times quicker than its autoxidation reaction [8].

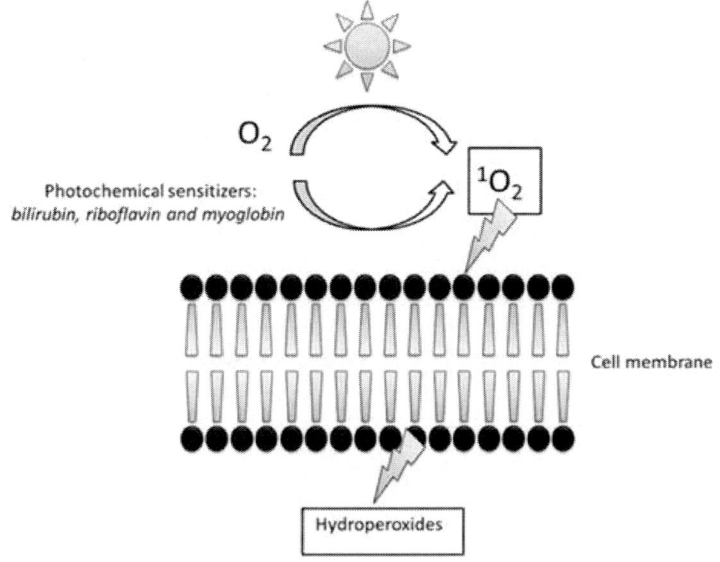


Figure 2. Photo-oxidation process of polyunsaturated fatty acids.

### Enzymatic Lipid Oxidation

Lipid peroxidation can be also catalyzed by oxidative enzymes such as lipoxygenase (LOX), cyclooxygenase (COX) and cytochrome P450 (CYP450) which yield very specific stereo- and regiospecific products (Figure 3).

Lipoxygenase (LOX) is an iron-containing enzyme that catalyzes the incorporation of molecular oxygen in a three consecutive step reaction to polyunsaturated fatty acids with a (Z,Z)-1,4-pentadiene structural unit and an activated methylene group between the two double bond (i.e., eicosapentaenoic acid, docosapentaenoic acid and arachidonic acid). This enzyme-catalyzed lipid oxidation produces lipid hydroperoxides such as leukotrienes, lipoxins, hepoxylins, hydroperoxyeicosatetraenoic acids (HpETEs), hydroxyeicosa-tetraenoic acid (HETE), hydroperoxy-octadecadienoic acid (HpODE) and hydroxyl octadecadienoic acid (HODE) [9, 10].

Cyclooxygenase (COX), also named as prostaglandin-endoperoxide synthase (PTGS) and which exists in two different isoforms COX-1 (constitutive) and COX-2 (inducible), catalyzes the oxidation of arachidonic acid (polyunsaturated omega-6 fatty acid 20:4) into prostaglandin H2 (PGH2). This enzymatic reaction consists of a stereoselective removal of the pro-S hydrogen at carbon 13 of the arachidonic acid and oxygenations with stereochemistry; there is a first oxygenation that occurs in the $11R$ configuration and a second oxygenation in the 15S configuration [11, 12].

Cytochrome P450 (CYP450) is a membrane-bound heme-containing metalloenzyme found in all mammalian cell types that catalyzes the NADPH-dependent oxidation of arachidonic acid to form epoxyeicosatrienoic acids (EETs) and hydroxyeicosatetraenoic acids (HETEs). As examples, we could highlight cytochromes P450 of the CYP4A gene subfamily which are responsible to produce HETEs, such as 20-HETE, and cytochromes P450 of the CYP2C and CYP2J gene subfamilies which are involved in the production of EETs in the kidney, heart, liver, brain and peripheral vasculature [13, 14].

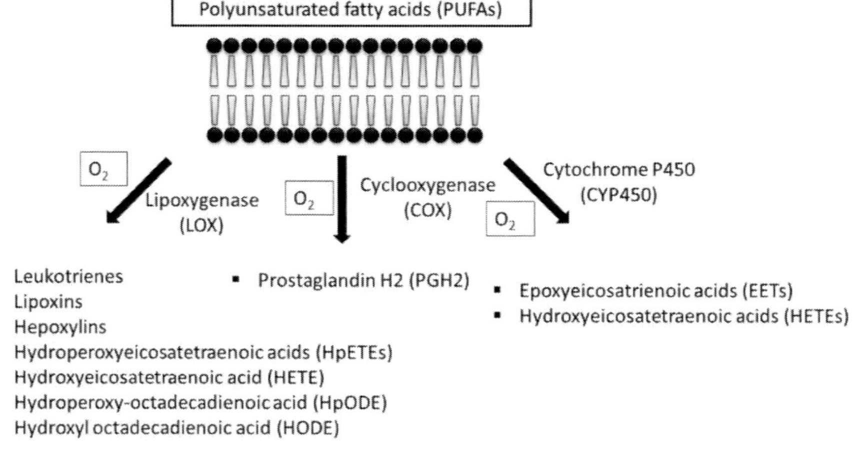


Figure 3. Enzymatic oxidation process of polyunsaturated fatty acids.

## 1.3. Malonyldialdehyde (MDA)

The process of lipid peroxidation of polyunsaturated fatty acids results in the formation of more than 200 different reactive aldehydes such as acrolein, 3-aminopropanal (3-AP), 4-oxononenal (4-ONE), 4-hydroxy 2-nonenal (4-HNE) and malondialdehyde (MDA) [15]. These reactive aldehydes, compared with free radicals, exhibit certain characteristics that significantly contribute to the cell damage: they have a significantly longer half-life and they are non-charged structures which allow them to move easily from a hydrophilic to a hydrophobic environment, reaching distant targets and extending the cellular damage [16]. These aldehydes are most reactive towards nucleophilic compounds; they bind to thiol and amino functional groups and modify free amino acids and amino acid residues in proteins, nucleobases of nucleic acids, and aminophospholipids by generating mostly irreversible covalent bonds (adducts). These events can result in impaired homeostatic balance, DNA and RNA damage, defective structural and regulating proteins, and undermining membrane integrity and function, among others [17, 18].

Malondialdehyde (MDA) is one of the most common and important end-product of lipid peroxidation. It is structurally characterized by containing a carbon-to-carbon double bond and a hydroxyl group (see chemical structure in Figure 4).

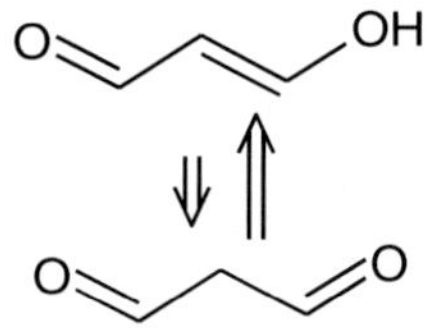

Figure 4. Chemical structure of malondialdehyde (MDA).

This product can be formed through two different pathways: (1) via autoxidation of arachidonic acid (AA) and larger polyunsaturated fatty acids (PUFAs) and subsequent β-scission of bicyclic endoperoxides; (2) during the enzymatic lipid oxidation of AA and PUFAs, malondyaldehyde is formed as a side product of thromboxane A2 (TXA2) and 12-1-hydroxy-5,8,10-heptadecatrienoic acid (HHT) biosynthesis. This three carbon dialdehyde is found in biological matrices into two forms: (1) as a free form (f-MDA) which is the chemically active form and it is widely used as an indicator of oxidative damage, and (2) as bound form to nucleophilic groups (SH or $NH_2$) of aminoacids, proteins, glycogen, and other biomolecules [4, 18, 19].

Various *in vitro*, *in vivo* and clinical studies have demonstrated that MDA can cause oxidative modifications of proteins by reacting with their amino groups of lysine, their sulfhydryl groups of cysteine, and their imidazole group of histidine to form cross-links and stable adducts on mammalian plasma, tissues and extracellular matrix (this process is known as carbonyl stress). The accumulation of this toxic carbonyl species, which may occur during ageing and in chronic diseases, such as neurodegenerative disorders, results in damaged and defective proteins in enzymes and in other proteins structures and functions and subsequently, impairment of multiple cellular functions [18, 20, 21].

*In vitro* studies have determined that MDA can interact with bovine serum albumin (BSA) inducing the generation of peroxides; this protein peroxide formation increases as MDA concentrations augmented and in presence of traces metal ions [22]. Moreover, it has been demonstrated in *in vivo* studies that MDA reacts with lysine groups of proteins to form MDA-lysine protein adducts in plasma and in hepatocytes of rats fed with a rich diet in iron [23]. In another study, as MDA-protein adducts are clinical useful markers of oxidative stress, Picaud et al. (2013) developed a non-invasive method to determine the formation of adducts between MDA and hemoglobin and measure their concentration in red blood cells in neonates [24]. Furthermore, it has been also confirmed that red blood cell lipids membranes of iron-deficient rats experience an increase in the lipid peroxidation product MDA; this metabolite react then with phospholipid and other high molecular weight protein complexes to form cross-linked MDA-proteins adducts which cause decreased erythrocyte deformability and consequently limiting its lifespan [25].

Besides the generation of protein adducts with malondialdehyde, there are other studies which have demonstrated that MDA react also with DNA bases to form adducts of deoxyguanosine ($M_1G$, $M_1dG$), deoxyadenosine ($M_1A$, OPdA), and deoxycytidine ($M_1C$, OPdC). These adducts have mutagenic and carcinogenic properties in bacteria and mammalian cells. In human beings, the major MDA-DNA adduct is pyrimido [1,2-alpha] purin-10(3H)-one ($M_1G$) which has been detected in blood white cells and in liver, pancreas and breast tissues; this major DNA adduct of the endogenous carcinogen malondialdehyde plays a key role in different kind of cancer [26].

The study of the mutagenicity in *Escherichia coli* of DNA adduct pyrimido [1,2-alpha] purin-10(3H)-one (M1G) revealed that the major mutation was two single base pair replacement (point mutations) M1G-->A and M1G-->T and this mutation may occur with a frequency of approximately 1% [27]. In another similar study, Marnett et al. (2003) demonstrated that pyrimido[1,2- alpha]purin-10(3H)-one (M1G) induces frameshift and base pair substitution mutations; particularly, the major MDA-DNA adduct induces -1 and -2 frameshift mutations when it is incorporated in a reiterated (CG)4 sequence in both *Escherichia coli* and COS-7 cells [28].

Few clinical studies have determined that the detection and quantification of MDA-DNA adducts help to improve the understanding of different types of cancer. For instance, a clinical study measured the levels of MDA-dA, MDAdG, and total MDA adducts in adjacent normal tissues of 51 breast cancer patients and compared with those in normal breast tissues of 28 non-cancer controls by using the highly sensitive 32P-postlabeling technique. Results revealed that DNA adducts induced by MDA are accumulated and present in higher concentrations in human breast cancer tissues than in human breast non-cancer tissues [29]. As another example, Bingham et al. (2002) detected in a prospective clinical trial the presence of malondialdehyde-deoxyguanosine (M(1)-dG) in human colorectal tissue by using a sensitive immunoblot blot assay. A positive association was found between DNA adducts and height and age in men and, between DNA adducts and saturated fat in women. On the other hand, an inverse association was found between DNA adducts and body mass index in men and, between DNA adducts and the ratio of polyunsaturated:saturated fatty acids [30]. Furthermore, it has been found that MDA–DNA adducts levels were higher in smokers patients with lung cancer and with GA and AA cyclin D1 (CCND1) genotypes. This study showed that the toxicity in lung cancer cellular was mediated by oxidative stress [31].

## 2. MALONDIALDEHYDE (MDA) AND ITS ROLE IN BIOLOGY OF NEURODEGENERATIVE DISEASES

Multiple studies have evidenced the potential role of malondialdehyde (MDA) in neurodegeneration and in central nervous disorders such as Alzheimer´s disease (AD), Parkinson´s disease (PD) and, amyotrophic lateral sclerosis (ALS). Assessing the concentrations of MDA in the central and in the peripheral compartments during neurodegenerative disorders is helpful to monitor the course of the disease and to measure how well treatment works [1-3].

Many *in vitro* models of neurodegenerative diseases (i.e., hydrogen-peroxide and 6-hydroxydopamine (6-OHDA) oxidative stress-induced cell toxicity) were investigated for MDA levels. Studies on human SH-SY5Y cells and PC12 rat cells, both cellular models widely employed for central nervous system disorders research, have revealed that different time incubations and concentrations of the previous mentioned toxic agents induced a significant MDA augmentation [32-36].

Different animal models of Alzheimer´s disease (AD), Parkinson´s disease (PD) and, amyotrophic lateral sclerosis (ALS), among other neurodegenerative disease models, have also investigated the increase in MDA levels. As example, in PD models, the injection of 6-OHDA for several weeks into the left striatum and into the substantia nigra of rats caused an elevation of MDA production in those areas of the brain compared to control rats [37, 38]. A pathogenic role of MDA has been also observed in AD animal models; the oxidative stress inductors scopolamine, amyloid β-protein (Aβ) and streptozotocin elevated MDA levels in mice and rats [39-42]. Oxidative stress and MDA also play a key role in amyotrophic lateral sclerosis (ALS) pathogenesis. Hence, studies on SOD(G93A) mouse animal model demonstrated high levels of this lipid peroxidation metabolite in tissues of spinal cord, brain, skeletal muscle and liver [43]. Moreover, high levels of MDA and MDA-protein adducts have been detected in the spinal cord of transgenic mice that overexpress a mutated human CuZn superoxide dismutase (SOD1) gene (gly93-->ala) [44].

Clinical studies and post-mortem studies on brain of people suffering from oxidative stress- related neurodegenerative diseases have also evidenced increases in MDA content. Elizabeth et al. (2014) on their study reported that the level of MDA was significantly increased and that the concentration of glutathione (GSH) and the activities of the antioxidant enzymes superoxide dismutase (SOD), catalase (CAT), glutathione peroxidase (GSH-Px) and glucose-6-phosphate dehydrogenase (G6PD) were significantly reduced in the erythrocytes of patients with Alzheimer´s disease and Parkinson´s disease [45]. Significant increases in the concentration of MDA have also been detected by high-performance liquid chromatography (HPLC) and thiobarbituric acid (TBA) test in human erythrocytes of subjects suffering from senile dementia of the Alzheimer type (SDAT) [46]. Regarding postmortem brains, there has been measured a high content of MDA in different regions of human brain suffering from a neurodegenerative pathology, such as in cerebellum, hippocampus and substantia nigra [47, 48].

## 3. METHODS FOR MDA DETECTION

Different experimental methods for MDA detection in cells and tissue samples have been developed (i.e., urine, plasma). One of the most common methods is the high-performance liquid chromatography (HPLC); in scientific literature, there have been compiled varieties in chromatographic method conditions of this separation technique such as different HPLC columns, flow rate and type of samples, among others. Focusing on HPLC methods for determining MDA in nervous system cells and brain samples, the following ones may be highlighted. Candan and Tuzmen (2008) developed and validated a HPLC method for the analysis of MDA in rat brain tissues using thiobarbituric acid (TBA) derivation reaction, new generation Phenomenex Gemini C18 and Hypersil ODS C18 columns with a mobile phase of 40:60 (vol/vol) methanol-$KH_2PO_4$ and fluorescence detection at 532 nm [49]. Kang et al. (2008) also employed a HPLC-assay based on TBA derivatization for determining MDA levels in the human endothelial ECV304 cell model (this cell line is commonly used for central nervous system studies); the chromatographic conditions for this HPLC method consist of a Synergi Hydro-RP column, a linear gradient of the mobile phase acetonitrile-ammonium acetate aqueous solution (10 mM, pH 6.8) and UV detection at 532 nm [50].

Besides chromatographic methods, there are other easy and rapid methods for screening and monitoring MDA content such as thiobarbituric acid reactive substances (TBARS) technique. TBARS assay has been widely employed to measure MDA concentration in neurodegenerative animal models (rats, mice) including chronic cerebral hypoperfusion- induced neurodegeneration [51], 6-hydroxydopamine induced PD [52], Abeta(1-40) and streptozotocin induced AD [53, 54], as some of multiple examples.

As another *in vitro* technique to measure MDA content highlights the immunocytochemical method developed by Hall et al. (1997). This research group employed this technique, which uses a rabbit-derived antibody raised against MDA-modified rabbit serum albumin (RSA), to localize and measure MDA content in central nervous system of transgenic familial amyotrophic lateral sclerosis mice and selectively vulnerable gerbil hippocampal CA1 region [55].

## 4. NATURAL PRODUCTS WITH BENEFICIAL EFFECTS ON LIPID PEROXIDATION IN OXIDATIVE STRESS-RELATED NEURODEGENERATIVE DISEASES

The current and most investigated therapeutic approach to neurodegenerative disorders implies the use of antioxidants. Antioxidants are considered as molecules capable to counteract the harmful effects derived from ROS-induced oxidation in biomolecules. Antioxidants may act through different mechanisms of action including scavenging of free radical and reactive species, up-regulation of endogenous antioxidant system and chelation of metal ions. They may be classified into two main categories based on their origin: endogenous and exogenous antioxidants. The endogenous antioxidant system comprises of several enzymes such as catalase, superoxide dismutase and glutathione peroxidase, among others, and a non-enzymatic component that includes molecules such as glutathione, uric acid and ubiquinol. Regarding exogenous antioxidants, one may highlight vitamins (i.e., vitamin C and E), polyphenols (i.e., flavonoids), carotenoids (i.e., zeaxanthine and lycopene), and isocyanates [56, 57].

Under conditions such as aging and neurodegenerative diseases, endogenous antioxidant defenses seem to be diminished (especially in brain) and inadequate to prevent from free radicals damage to DNA, lipids, proteins, and other biomolecules. Oxidative stress-mediated cellular injury has actually been considered as a major cause of neurodegenerative diseases. Therefore, the discovery, development and the use of different exogenous antioxidants that might ameliorate neural injury by oxidative stress have been increased during the last years. In this context, natural antioxidants have shown to be efficient inhibitors of the oxidative process, and they are considered as a promising therapeutic option [58].

Concerning brain, it presents special characteristics that make it more vulnerable to oxidative stress than other tissues, including the high content of unsaturated lipids (80%) and autooxidisable molecules (i.e., dopamine and serotonine) that are endogenous sources of ROS [59]. Hence, lipid peroxidation is considered as one of the main organic expressions by which oxidative stress exerts harmful effects in nervous tissue [60]. Lipid peroxidation can be prevented via reducing the formation of free radicals by: (a) destroying the free radicals that are already formed, (b) supplying a competitive substrate for unsaturated lipids in the membrane, and (c) accelerating the repair mechanism of damaged cell membrane. In cases of reduced or impaired *in vivo* antioxidant defense, exogenously administered antioxidants may be helpful to overcome the oxidative damage caused by free radicals [61].

Numerous compounds from natural origin have demonstrated to exert valuable properties in reducing lipid peroxidation processes, thus attenuating MDA formation and derived cellular damages in neurodegeneration models. Taking into consideration all points discussed so far, a compound that inhibits the chain reactions or reverses any factor enhancing lipid peroxidation in a neurodegenerative disorder will be of potential application as a therapeutic agent. In the paragraphs below, we aim to collect extensive information related to the capacities of natural compounds and extracts to display neuroprotective effects by mechanisms that include the reduction in MDA production; we collect recent research data on representative molecules from the main groups of structures (see Figure 5 and Table 1), as well as extracts of relevant plants (see Table 2).

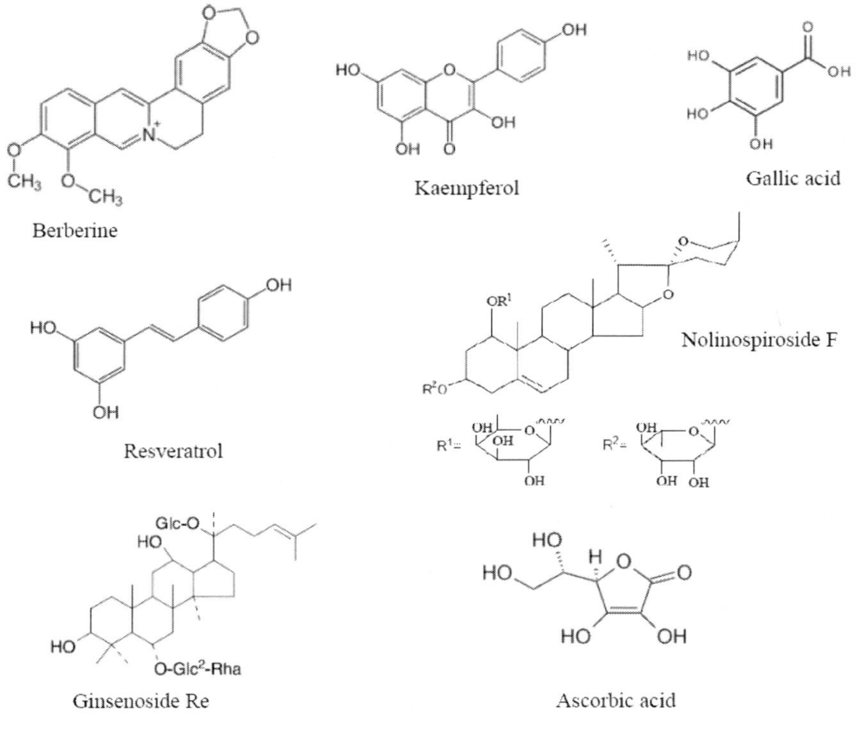

Berberine

Kaempferol

Gallic acid

Resveratrol

Nolinospiroside F

Ginsenoside Re

Ascorbic acid

Figure 5. Representative natural compound with interest as antioxidants against lipid peroxidation.

An abnormally elevated lipid peroxidation has been largely documented in brains of Alzheimer´s disease (AD) patients, and it is by far the neurodegenerative disorder that has attracted more interest and studies, both *in vitro* and *in vivo*, regarding the potential application of natural antioxidants for its management [62]. Choi et al. (2001) evaluated the mechanisms of the *in vitro* neuroprotective effects exerted by the green tea flavonoid (-)-

epigallocatechin gallate (EGCG), on an amyloid ß (Aß)- induced AD model of cultured rat hippocampal neurons. Through the thiobarbituric acid assay, it was demonstrated that a co-treatment of cells with 10 µM EGCG prevented from the increase in lipid peroxidation, as evidenced by a reduction of MDA levels to a basal range [63]. Another flavonoid, naringenin, improved learning and memory abilities on a rat model of AD induced by streptozotozin; it limited *in vivo* the extent of lipid peroxidation when co-administrated intragastrically for 3 weeks, by significantly reducing MDA levels in comparison with non-treated rats [64].

Concerning plant extracts, the known *Ginkgo biloba* arises as an important source of interesting natural antioxidants and one of the best studied species in terms of lipid peroxidation inhibition. Its standardized extract EGb761, composed of flavoglycosides, terpene lactones and ginkgolic acid, displayed protective actions on an Aβ- induced rat model of AD; EGb761, when injected intraperitoneally at a concentration of 20 mg/kg for 15 days, was able to ameliorate oxidative stress markers, including a reduction of hippocampal levels of MDA [65]. Similarly, an ethanol extract of *Melissa officinalis* exerted a lowering effect of lipid peroxidation *in vitro* on Aβ(25-35)- induced model of AD in rat pheochromocytoma PC12 cells. Pretreatment of cells with 10 µg/mL of the extract or 1 µg/mL of its acidic fraction significantly attenuated the MDA production to values comparable with those of control cells; such action might be attributed to the main constituents of the extract, that include polyphenols such as rosmarinic acid, caffeic acid and ursolic acid [66]. A mixture of several timosaponins (B2, E1 and B) extracted from the plant *Anemarrhena aspholeloides*, a traditional Chinese herbal drug, showed interesting effects on learning and memory ability of rats with dementia induced by lateral cerebral ventricular injection of Aβ peptide. Such treatment elevated lipid peroxidation in brain and, in the contrary, intragastric administration of timosaponins at three different doses (50, 100 and 200 mg/kg/day for 14 days) upgraded antioxidant capacity and reduced MDA levels in brain tissue from 5.95 to 4.94 - 4.67 nmols MDA/mg protein [67].

It is also remarkable the beneficial effects on AD-related lipid peroxidation displayed by some inhaled volatile essential oils on disease models. With this regard, Cioanca et al. (2013) evaluated the cognitive-enhancing properties exerted by the essential oil of the traditionally used *Coriander sativum* in Aβ(1-42)- induced rat model of AD. They proved that the linalool-rich coriander volatile oil, when inhaled by rats at 1% and 3%, daily, for 21 days, improved spatial memory performance as well as oxidative stress markers; it was correlated to an attenuation in the increased MDA levels in hippocampus, that were even lower than basal levels[68]. On an analogous model of AD, perfume stimulating olfaction with volatile oil of *Acorus Gramineus* ameliorated the learning-memory ability of rats, accompanied by a significant reduction of MDA levels in brain, from 4.51 to 2.87 nmols/mg protein [69].

After Alzheimer´s disease, Parkinson´s disease (PD) is the most prevalent neurodegenerative disorder and also attracts much research on the present issue. Kaempferol, an ubiquitous naturally occurring flavonoid, was shown to protect against the deleterious effects of 1-methyl-4-phenyl-1,2,3,6-tetrahydropyridine (MPTP) in mice. Pretreatment with kaempferol (50 and 100 mg/kg, orally, for 14 days) improved motor coordination, raised striatal dopamine and amended redox status, as evidenced by a low content of MDA (compared with mice treated with MPTP alone) which was reduced from 13.23 to 9.70 and 8.75 nmols/mg protein [70]. The phenolic compound gallic acid, present together with its derivatives in grapes, berries and many other fruits, exerted neuroprotection in a 6-hydroxydopamine (6-OHDA)- induced PD model in rats. Besides motor memory improvements, oral gallic acid pretreatment markedly decreased MDA levels in hippocampus and striatum of rats [71]. The previously mentioned *Ginkgo biloba* also displayed protective effects against PD-like symptoms. An extract obtained from its leaves was administered intraperitoneally in rats, at a concentration of 100 mg/kg/day for 19 days, prior to microinjection of MPTP in the substantia nigra; such treatment was able to significantly reduce lipid peroxidation in the substantia nigra of rats [72]. In the same way, Hritcu et al. (2011) have recently demonstrated neuroprotective potential for a methanol extract of *Hibiscus asper*, a plant from Malvaceae family traditionally used in tropical Africa medicine. In 6-OHDA- lesioned rats, chronic administration of its methanolic extract, mainly containing flavonoids, polyphenols and saponines, showed potent antioxidant actions including reduced MDA levels in rat temporal lobe homogenates (lower than those of control rats), as evidenced by TBARS measurements [73]. In addition, the investigation conducted by Yadav et al. (2013) deserves to be pointed out; they developed a special model of PD through a chronic exposure of mice to paraquat in order to assess the activity of the traditional herbal medicine *Mucuna puriens*, which contains many micronutrients such as amino acids, Zn, Se, carbohydrates and various plant alkaloids. Treatment with an aqueous seed extract of the plant improved behavioral abnormalities and prominently decreased MDA levels in nigrostriatal regions, when compared to untreated Parkinsonian mouse brain [74].

**Table 1. Beneficial effects of natural compounds on lipid peroxidation in models of neurodegeneration**

| Compound | Origin | Model of neurodegeneration | Method | Effects observed | Reference |
|---|---|---|---|---|---|
| **Alkaloids** | | | | | |
| Berberine | Comercial (Chengdu 3rd Pharmaceutical) | Aluminium- induced rat model of neurodegeneration (*in vivo*) | Comercial MDA assay kit | Berberine (100 mg/kg) significantly blunted the increase in MDA, from 1.93 to 0.72 nmol MDA/mg protein | [77] |
| Huperzine B | *Huperzia serrata* | $H_2O_2$- induced toxicity in rat pheochromocytoma line PC12 (*in vitro*) | Thiobarbituric acid method | Pretreatments with huperzine B (10-100 μM) decreased the level of MDA in around 25% when compared with $H_2O_2$- treated cells | [78] |
| **Carotenoids** | | | | | |
| Crocin | Comercial (Fluka) | Streptozotocin- induced rat neurodegeneration model (*in vivo*) | TBARS measurements | Crocin (100 mg/kg, p.o.) markedly decreased MDA levels in hippocampus and cerebral cortex of rats | [79] |
| **Flavonoids** | | | | | |
| (-)-Epigallocatechin gallate (EGCG) | Comercial (Sigma Chemical Co) | Aβ(25-35)- induced model of AD in primary hippocampal rat neurons (*in vitro*) | TBARS measurements | Cotreatments of cells with EGCG could prevent the increase in MDA levels | [63] |
| Kaempferol | Comercial (Sigma-Aldrich) | MPTP- induced PD model in mouse (*in vivo*) | Comercial MDA assay kit | Kaempferol (50 and 100 mg/kg, p.o. for 14 days) partially reduced the content of MDA compared with MPTP-treated mice | [70] |

| Compound | Origin | Model of neurodegeneration | Method | Effects observed | Reference |
|---|---|---|---|---|---|
| **Flavonoids** | | | | | |
| Naringenin | - | Streptozotocin- induced rat AD model (*in vivo*) | Colorimetric method | Naringenin (intragastric, for 3 weeks) significantly decreased MDA levels | [64] |
| **Phenolic compounds** | | | | | |
| Curcumin | Comercial (Sigma-Aldrich) | 3-NP- induced model of HD in rats (*in vivo*) | TBARS measurements | Chronic curcumin administration (20 and 50 mg/kg, p.o. for 8 days) significantly prevented the increase in MDA levels | [75] |
| Gallic acid | Comercial (Sigma-Aldrich) | 6-OHDA- induced rat model of PD (*in vivo*) | TBARS measurements | Gallic acid pretreatment (50, 100 and 200 mg/kg, p.o. for 10 days) decreased MDA levels in hippocampus and striatum | [71] |
| Macranthoin G (MCG) | *Eucommia ulmoides* | $H_2O_2$- induced toxicity in rat pheochromocytoma line PC12 (*in vitro*) | TBARS measurements | Pretreatment with MCG (6.25-50 μM) significantly decreased MDA content in a dose-dependet manner | [83] |
| Paeonol | Comercial (Xuancheng Baicao Plants) | D-galactose- induced neurotoxicity in mice (*in vivo*) | Comercial MDA assay kit | Paeonol (100 mg/kg/day) decreased MDA content almost to control levels | [91] |
| Resveratrol | Comercial (Sigma Chemical Co.) | $H_2O_2$- induced oxidative stress in rat cortical cell culture (*in vitro*) | TBARS measurements | Pretreatment with 10 μM resveratrol and 1 μM resveratrol in nanoparticles significantly decreased MDA levels | [82] |

**Table 1. (Continued)**

| Compound | Origin | Model of neurodegeneration | Method | Effects observed | Reference |
|---|---|---|---|---|---|
| **Steroids** | | | | | |
| Cholesterol | Mussel sterols extract | $H_2O_2$- induced K6001 yeast strain model | Comercial MDA assay kit | Cholesterol 5 or 3 µM decrease MDA level under oxidative stress condition | [80] |
| Nolinospiroside F | *Ophiopogon japonicus* | $H_2O_2$- induced K6001 yeast strain model | Comercial MDA assay kit | Doses of 1, 3 and 10 µM decrease the level of MDA | [81] |
| Terpenoids | | | | | |
| Ginsenoside Re | Comercial (Shenyang Pharmaceutical University) | Cerebral ischemia-reperfusion model in rats *(in vivo)* | Comercial MDA assay kit | Pretreatment with Re (10 and 20 mg/ kg, p.o. for 7 days) significantly reduced MDA levels in a dose-dependent manner | [92] |
| Ginsenoside Rg2 | - | Glutamate- induced neurotoxicity in rat pheochromocytoma PC12 cell model *(in vitro)* | Comercial MDA assay kit | Pretreatments with Rg2 (0.05, 0.1, 0.2 mM) markedly decreased MDA levels in the cells | [85] |
| Linearol Sidol | *Sideritis* spp. | $H_2O_2$- induced oxidative stress in rat PC12 cell model *(in vitro)* | HPLC analysis | Pretreatment with linearol and sidol (5 and 10 µM, 24 h) significantly decreased MDA levels | [90] |
| Pinusolide and 15-methoxypinusolidic acid | *Biota orientalis* (L.) Endl. (Cupressaceae) leaves | Staurosporine- induced rat primary cortical neurons apoptosis *(in vitro)* | TBARS measurements | At concentrations of 5 µM, they significantly decrease MDA levels | [93] |

| Compound | Origin | Model of neurodegeneration | Method | Effects observed | Reference |
|---|---|---|---|---|---|
| **Other structures** | | | | | |
| Ascorbic acid (dihydroxyfuranone) | - | Gentamicin- induced lipid peroxidation model in rabbit (*in vivo*) | TBARS measurements | Ascorbic acid (40 mg/kg, i.m., 24 h) reduced the previously increased MDA content in blood | [61] |
| Parathelypteriside (stilbenoid compound) | *Parathelypteris glanduligera* (kze.) ching | D-galactose- induced neurotoxicity in mice (*in vivo*) | TBARS measurements | Injection of PG (5 or 10 mg/kg/day, i.p.) for 2 weeks reduced the increased levels of MDA | [84] |
| Tetramethylpyrazine (TMP) | Comercial (Sigma-Aldrich) | Kainate- induced toxicity in rats (*in vivo*) | TBARS measurements | TMP (50 μM) substantially decreased MDA to normal levels in primary cultures of hippocampal neurons and rat hippocampus | [94] |

MDA: malondialdehyde; TBARS: thiobarbituric acid reactive substances; p.o.: per os (orally); i.p.: intraperitoneally; i.m.: intramuscular; i.g.: intragastrically; 6-OHDA: 6-hydroxidopamine; 3-NP: 3-nitropropionic acid; HD: Huntintong´s disease; AD: Alzheimer´s disease; PD: Parkinson´s disease; MPTP: 1-methyl-4-phenyl-1,2,3,6-tetrahydropyridine.

**Table 2. Beneficial effects of plant extracts on lipid peroxidation in models of neurodegeneration**

| Extract | Composition | Neurodegeneration model | Method | Parameters measured and effects observed | Reference |
|---|---|---|---|---|---|
| *Acorus gramineus* volatile oil | Not studied | Aβ(1-40)- induced model of AD in rats (*in vivo*) | TBARS measurements | Inhaled volatile oil (5 mL, 60 min/day for 42 days) was able to reverse the increase in MDA level in brain tissues | [69] |
| *Allium sativum* (aged garlic) ethanol extract | S-allyl-cysteine | $FeSO_4$- induced lipid peroxidation in mice brain homogenate (*in vitro*) | TBARS measurements | Ethyl acetate fraction of the extract (200 µg/ml) suppressed MDA levels in more than 50% | [86] |
| *Anemarrhena aspholeloides* extracted timosaponins | Timosaponin B2, timosaponin E1 and timosaponin B | Aβ(1-42)- induced rat model of dementia (*in vivo*) | TBARS measurements | Timosaponins (50, 100 and 200 mg/kg), i.g. for 14 days) reduced MDA levels in brain tissue | [67] |
| *Bacopa monnieri* aqueous extract | Saponin glycosides: bacoside A3, bacopaside II, bacopasaponin C isomer, bacopasaponin C and bacopaside I | Aβ(25-35)- induced model of AD in rat primary cortical neurons (*in vitro*) | TBARS measurements | At concentrations higher than 150 µg/ml of the extract, it almost completely blocked (80%) the generation of lipid peroxide products | [95] |
| *Centella asiatica* aqueous extract | Asiaticoside, medecassocide and brahmic acid | Streptozotocin- induced AD model in rat (*in vivo*) | TBARS measurements | Extract (200 and 300 mg/kg, for 21 days) significant decrease in MDA levels | [96] |
| *Coptis chinensis* total base (CTB) | Berberine (87.6%), palmatine (2.7%), jatrorrizhine (3.6%) and coptisine (5.2%) | Aluminium- induced rat model of neurodegeneration (*in vivo*) | Comercial MDA assay kit | CTB (110 mg/kg) significantly blunted the increase in MDA, from 1.93 to 0.44 nmol MDA/mg protein | [77] |

| Extract | Composition | Neurodegeneration model | Method | Parameters measured and effects observed | Reference |
|---|---|---|---|---|---|
| *Coriandrum sativum* volatile oil | Linalool | Aβ(1-42)- induced rat model of AD (*in vivo*) | TBARS measurements | Inhaled volatile oil (1% or 3% daily for 21 days) attenuated the increase in hippocampal MDA level | [68] |
| *Ginkgo biloba* leaves extract (GBE) | 24% Ginkgo flavonoid glycosides, 6% terpene lactone | Trimethyltin- induced neurotoxicity in rats | Linoleic acid peroxidation method | GBE (70 mg/kg, i.p. for 14 days) partially attenuated the induced increase in MDA levels | [97] |
| *Ginkgo biloba* leaves extract (GBE) | | MPTP- induced model of PD in rats (*in vivo*) | TBARS measurements | GBE (50 and 100 mg/kg, i.p. for 20 days) significantly decrease MDA levels in substantia nigra of rats | [72] |
| *Ginkgo biloba* standardized extract (EGb 761) | Flavoglycosides, terpene lactones and ginkgolic acid | 3-Nitropropionic acid- induced HD rat model (*in vivo*) | TBARS measurements | EGb 761 (100 mg/kg, i.p. for 15 days) pretreatment attenuated the level of striatal MDA | [76] |
| *Ginkgo biloba* standardized extract (EGb 761) | | Aβ(1-25)- induced rat model of AD (*in vivo*) | Comercial MDA assay kit | EGb 761 (20 mg/kg, i.p. for 15 days) pretreatment attenuated the level of hippocampal MDA | [65] |
| *Hibiscus asper* leaves methanol extract | Flavonoids, polyphenols and saponines | 6-OHDA- induced model of PD in rats (*in vivo*) | TBARS measurements | Methanolic extract (50 and 100 mg/kg/day, i.p. for 7 days) significantly reduced MDA levels in temporal lobe homogenates to lower values than in control rats | [73] |
| Extract | Composition | Neurodegeneration model | Method | Parameters measured and effects observed | Reference |
| *Lavandula* spp. essential oils from | Linalool, linalyl acetate, terpinen-4-ol, lavandulyl acetate, camphor and borneol | Scopolamine- induced dementia model in rats (*in vivo*) | TBARS measurements | Inhaled essential oils (200 µl, 60 min/day for 7 days) significantly diminished MDA levels in temporal lobe homogenates (lower than in control rats) | [87] |

**Table 2. (Continued)**

| Extract | Composition | Neurodegeneration model | Method | Parameters measured and effects observed | Reference |
|---|---|---|---|---|---|
| *Melissa officinalis* ethanol extract | Rosmarinic acid, caffeic acid and ursolic acid | Aβ(25-35)- induced model of AD in rat pheochromocytoma PC12 cells (*in vitro*) | TBARS measurements | Pretreatment of cells with 10 µg/mL of the extract and 1 µg/mL of its acidic fraction significantly attenuated the Aβ-induced MDA production | [66] |
| *Moringa oleifera* leaves hidroalcoholic extract | ß-carotene, vitamin C and flavonoids | Cholinotoxin- induced model of age-related dementia in rats (*in vivo*) | TBARS measurements | The extract (100 and 200 mg/kg, p.o., 14 days) reverted the increase in MDA generation to basal levels | [98] |
| *Mucuna pruriens* seed extract | Alkaloids, aminoacids (L-Dopa), Zn and Se | Paraquat- induced model of PD in mice (*in vivo*) | TBARS measurements | The extract (100 mg/kg, p.o. for 3, 6 and 9 weeks) significantly reduced MDA levels in the nigrostriatal region of brains | [74] |
| *Scutellaria baicalensis* total flavonoids, extracted from stems and leaves (SSF) | Not studied | $H_2O_2$- induced toxicity in rat pheochromocytoma line PC12 (*in vitro*) | TBARS measurements | Pretreatment with SSF (18–76 µg/mL) notably lowered the MDA level | [88] |
| *Vitis amurensis* methanol extract from leaves and stems | (+)-ampelopsin A, γ-2-viniferin, and trans-ε-viniferin | Cerebral ischemic damage induced in rats (*in vivo*) | TBARS measurements | Orally administered *V. amurensis* (25–100 mg/kg) attenuated the increase in lipid peroxidation | [99] |
| *Zingiber officinale* root extract (GRE) | Gingerol and shogaol | Aβ and aluminium-induced model of AD in rats (*in vivo*) | Comercial ELISA kit | GRE (4 g/kg, i.g. for 35 days) significantly decreased levels of serum MDA | [89] |

| Extract | Composition | Neurodegeneration model | Method | Parameters measured and effects observed | Reference |
|---|---|---|---|---|---|
| *Zingiber officinale* var. Rubra and *Zingiber officinale* Roscoe aqueous extracts | Volatile oils and phenol compounds as gingerols, sesquiterpenoids and shogaols | $Fe^{2+}$- induced lipid peroxidation model in rat brain homogenates (*in vitro*) | TBARS measurements | Water extracts (0.78–3.13 mg/mL) caused a dose-dependent significant decrease in the MDA content | [100] |

MDA: malondialdehyde; TBARS: thiobarbituric acid reactive substances; p.o.: per os (orally); i.p.: intraperitoneally; i.m.: intramuscular; i.g.: intragastrically; 6-OHDA: 6-hydroxidopamine; 3-NP: 3-nitropropionic acid; HD: Huntintong´s disease; AD: Alzheimer´s disease; PD: Parkinson´s disease; MPTP: 1-methyl-4-phenyl-1,2,3,6-tetrahydropyridine.

Some natural products have been studied in models of other neurodegenerative diseases. For instance, phenolic compound curcumin has been assayed *in vivo* in a model of Huntintong´s disease induced by 3-nitropropionic acid (3-NP), an irreversible inhibitor of succinate dehydrogenase enzyme. After administration of 3-NP, rats showed declined motor function and memory, and changes in oxidative stress markers; however, chronic treatment with curcumin (10, 20 and 50 mg/kg/day, orally, for 8 days) was able to revert those signs dose-dependently, including a prevention of the increase in cerebral MDA levels [75]. Also, the effects of the *Ginkgo biloba* extract Egb761 have been recently evaluated in such model of HD. Neurodegeneration induced by 3-NP in rats was partially reversed by a pretreatment with Egb761, which resulted in amelioration of locomotor behavior and biochemical parameters; it was able to decrease the upregulated levels of striatal MDA, and stabilize them in levels similar to control rats [76].

On the other hand, lipid peroxidation is a common marker used for measuring oxidative stress and cellular damage in other models of general neurodegeneration and neurotoxicity, not necessarily identified with a certain disease. In recent times, many natural compounds and plant extracts has been investigated in relation to their capacities to protect against several neurotoxic agents, which typically produce an increase in lipid peroxidation in *in vivo* and *in vitro* models.

With regard to isolated compounds, the alkaloid berberine, present in the rhizome of *Coptis chinensis*, appears to be a promising neuroprotective agent. Berberine was able to attenuate the brain injury induced by aluminium in a rat model of neurodegeneration; apart from stabilizing several enzyme activities, co-treatment with the alkaloid significantly blunted the increase in MDA occurring at the hippocampus of rats, from 1.93 to 0.72 nmols MDA/mg protein [77]. Another plant alkaloid, huperzine B, isolated from *Huperzia serrata*, diminished the hydrogen peroxide- induced oxidative stress and toxicity in PC12 cells. Pretreatments of cells with huperzine B (10-100 mM) prior to $H_2O_2$ exposure significantly elevated cell survival, antioxidant enzyme activities and decreased the level of MDA [78]. The carotenoid crocin displayed protective effects against the cognitive impairment, associated with free radical generation, due to intracerebroventricular injection of streptozotocin in rats. When orally administered at a dose of 100 mg/kg for 21 days, crocin markedly decreased MDA levels in hippocampus and cerebral cortex of rats, as measured through TBARS assay [79]. Throughout their research on natural steroids, Sun et al. (2013; 2014) demonstrated *in vitro* neuroprotective effects via antioxidant potential for cholesterol, isolated from mussels, and nolinospiroside F, a saponin obtained from the plant *Ophiopogon japonicus*. In a model of $H_2O_2$- induced oxidative stress in K6001 yeast strain, treatment with different doses of both compounds extended lifespam of yeasts and avoided the enhancement of MDA production [80, 81].

In this sense, many authors have employed hydrogen peroxide as a neurotoxic agent, due to its capacity to induce oxidative stress and augment lipid peroxidation, for evaluating the *in vitro* neuroprotective activities of many natural compounds. For example, the popular antioxidant resveratrol has been studied regarding its capacity to modulate lipid peroxidation in neurodegeneration. Since application of this phenolic compound is limited for its insolubility and short half-time, Lu et al. (2013) developed nanoparticle-based delivery systems in order to improve those properties and tested them in a $H_2O_2$- induced model of neurodegeneration *in vitro*. They demonstrated that a pretreatment of rat cortical cells in culture with 1 µM resveratrol in nanoparticles significantly decreased MDA levels in more than half of model value; higher concentration (10 µM) of free resveratrol was needed to get such effect [82]. Macranthoin G, another phenolic compound isolated from *Eucommia ulmoides*, has exhibited neuroprotective actions against $H_2O_2$-induced apoptosis in PC12 cells. Inhibition of lipid peroxidation might be a plausible mechanism for such effect, as a pretreatment with 6.25-50 µM macranthoin G reduced cellular MDA content in a dose-dependent manner [83].

Other models of neurotoxicity include the use of noxious substances such as D-galactose and glutamate. The stilbenoid compound parathelypteriside, extracted from *Parathelypteris glanduligera* has recently been shown to attenuate cognition deficits in D-galactose- treated mice via up-regulation of antioxidant capacity; as evidenced by TBARS measurements, intraperitoneal injection of parathelypteriside (5 and 10 mg/kg/day for 2 weeks) reduced the increased levels of MDA in the hippocampus of D-galactose- treated animals from 5.10 to 4.88 and 4.17 nmols MDA/mg protein, respectively [84]. The terpenoid compound ginsenoside Rg2 also displayed protective effects against glutamate-induced neurotoxicity in PC12 cells. When the cells were co-incubated with glutamate 1 mM and ginsenoside Rg2 (0.05, 0.1, 0.2 mM) for 24 h, lipid peroxidation products, expressed as MDA, were significantly reduced in comparison with cells only treated with glutamate (with 200-250 % control values) [85].

If we focus on plant extracts, it is remarkable the study which proved that aged garlic ethanol extracts (*Allium sativum*) exerted ameliorating effects against *in vitro* Aβ-induced neurotoxicity and cognitive impairment. Their capacity to inhibit lipid peroxidation was tested in a model of $FeSO_4$- induced lipid peroxidation in mice brain homogenate; effective action against such process was acknowledge to the ethyl acetate fraction of the extract (200 µg/ml), since it suppressed MDA levels in more than 50% [86]. Hancianu et al. (2013) evidenced neuroprotective effects for essential oil of *Lavandula* spp. on scopolamine-induced dementia via anti-oxidative activities in rats. Subacute exposures (200 µl, 60 min/day for 7 days) of rats to inhaled lavender oil decreased the levels of lipid peroxidation in rat temporal lobe homogenates, presenting contents of MDA even lower than those in control animals [87]. Moreover, a mixture of flavonoids extracted from the stems and leaves of *Scutellaria baicalensis* prevented from the oxidative injury produced by hydrogen peroxide in PC12 cells. Through TBARS measurements, it was demonstrated that pretreatment of cells with total flavonoids (18-76 µg/mL) notably lowered the MDA level in a dose-dependent manner, reducing it from 2.98 to 0.71 nmols MDA/mL at the highest concentration the extract; this antioxidant effect may contribute to the potential clinical efficacy of this plant in the treatment of neurodegenerative diseases [88].

Finally, most of the studies measuring the effects of natural products on lipid peroxidation in neurodegeneration models estimate the concentration of MDA through the use of thiobarbituric acid assay; this assay is based on a colorimetric reaction and spectrophotometric determinations. However, some recent works quantify the degree of lipid peroxidation by using other methods. For instance, Zang et al. (2013) used a commercially available enzyme-linked immunoassay (ELISA) kit for calculating the concentration of MDA in rat serum. They demonstrated, together with the protective effects of ginger root extract on an *in vivo* model of AD, the capacity of the extract (4 g/kg/day for 35 days) to significantly reduce the levels of the lipid peroxidation marker [89]. Meanwhile, González-Burgos et al. (2013) assessed the neuroprotective properties of terpenoid compounds (linearol and sidol) in an *in vitro* model of $H_2O_2$- induced oxidative injury, and their antioxidant action on lipid peroxidation was evaluated through a more sensitive HPLC method; by comparing retention time of standard MDA and that in the cells samples, it was proved that the natural terpenoids could reduce the extent in lipid peroxidation [90].

# REFERENCES


[1]     Z. Chen, C. Zhong. *Neurosci. Bull.* 30, 2 (2014).

[2]     J. Li, W. O, W. Li, Z.G. Jiang, H.A. Ghanbari. *Int. J. Mol. Sci.* 14, 12 (2013).

[3]     V. Dias, E. Junn, M.M. Mouradian. *J. Parkinsons Dis.* 3, 4 (2013).

[4]     A. Ayala, M.F. Muñoz, S. Argüelles. *Oxid. Med. Cell Longev.* 2014 (2014).

[5]     M. Repetto, J. Semprine, A. Boveris. In: Catala A. (ed) *Lipid peroxidation*. InTech, Rijeka, chapter 1, pp. 1-28.

[6]     H.S. El-Beltagi, H.I. Mohamed. *Not. Bot. Horti.* Agrobo. 41, 1 (2013).

[7]     P.M. Yasaei, G.C. Yang, C.R. Warner, D.H. Daniels, Y. Ku. *J. Am. Oil Chem. Soc.* 73, 9 (1996).

[8]     E.N. Frankel, W.E. Neff, T.R. Bessler. *Lipids* 14 (1979).

[9]     S. Yamamoto. *Free Radic. Biol. Med.* 10, 2 (1991).

[10]    M. Maccarrone, G. Melino, A. Finazzi-Agrò. *Cell Death Differ.* 8, 8 (2001).

[11]    L.J. Marnett, S.W. Rowlinson, D.C. Goodwin, A.S. Kalgutkar, C.A. Lanzo. *J. Biol. Chem.* 274, 33 (1999).

[12]    C. Schneider, D.A. Pratt, N.A. Porter, A.R. Brash. *Chem. Biol.* 14, 5 (2007).

[13]    M.L. Edin, Z. Wang, J.A. Bradbury, J.P. Graves, F.B. Lih, L.M. DeGraff, J.F. Foley, R. Torphy, O.K. Ronnekleiv, K.B. Tomer, C.R. Lee, D.C. Zeldin. *FASEB J.* 25, 11 (2011).

[14]    J.K. Chen, J. Capdevila, R.C. Harris. *Mol. Cell Biol.* 21, 18 (2001).

[15]    S. Singh, C. Brocker, V. Koppaka, Y. Chen, B.C. Jackson, A. Matsumoto, D.C. Thompson, V. Vasiliou. *Free Radic. Biol. Med.* 56 (2013).

[16]    A.J. Hulbert, R. Pamplona, R. Buffenstein, W.A. Buttemer. *Physiol. Rev.* 87, 4 (2007).

[17]    D. Matveychuk, S.M. Dursun, P.L. Wood, G.B. Baker. *Bull. Clin. Psychopharmacol.* 21, 4 (2011).

[18]    H. Esterbauer, R.J. Schaur, H. Zollner. *Free Radic. Biol. Med.* 11, 1 (1991).

[19]    F. Bamonti, C. Novembrino, S. Ippolito, E. Soresi, A. Ciani, S. Lonati, E. Scurati-Manzoni, G. Cighetti. *Clin. Chem. Lab. Med.* 44, 4 (2006).

[20]    H. Esterbauer, M. Dieber-Rotheneder, G. Waeg, G. Striegl, G. Jurgens. *Chem. Res. Toxicol.* 3, 2 (1990).

[21]    H. Esterbauer. *Am. J. Clin. Nutr.* 57, 5 (1993).

[22] N. Traverso, S. Menini, E.P. Maineri, S. Patriarca, P. Odetti, D. Cottalasso, U.M. Marinari, M.A. Pronzato. *J. Gerontol. A. Biol. Sci. Med. Sci.* 59, 9 (2004).

[23] K. Houglum, M. Filip, J.L. Witztum, M. Chojkier. *J. Clin. Invest.* 86, 6 (1990).

[24] C. Cipierre, S. Haÿs, D. Maucort-Boulch, J.P. Steghens, J.C. Picaud. *Oxid. Med. Cell Longev.* 2013 (2013).

[25] S.K. Jain, R. Yip, R.M. Hoesch, A.K. Pramanik, P.R. Dallman, S.B. Shohet. *Am. J. Clin. Nutr.* 37, 1 (1983).

[26] L.J. Marnett. *Mutat. Res.* 424, 1-2 (1999).

[27] S.P. Fink, G.R. Reddy, L.J. Marnett. *Proc. Natl. Acad. Sci. U.S.A.* 94, 16 (1997).

[28] L.A. VanderVeen, M.F. Hashim, Y. Shyr, L.J. Marnett. *Proc. Natl. Acad. Sci. U.S.A.* 100, 24 (2003).

[29] M. Wang, K. Dhingra, W.N. Hittelman, J.G. Liehr, M. de Andrade, D. Li. *Cancer Epidemiol. Biomarkers Prev.* 5, 9 (1996).

[30] C. Leuratti, M.A. Watson, E.J. Deag, A. Welch, R. Singh, E. Gottschalg, L.J. Marnett, W. Atkin, N.E. Day, D.E. Shuker, S.A. Bingham. *Cancer Epidemiol. Biomarkers Prev.* 11, 3 (2002).

[31] A. Munnia, S. Bonassi, A. Verna, R. Quaglia, D. Pelucco, M. Ceppi, M. Neri, M. Buratti, E. Taioli, S. Garte, M. Peluso. *Free Radic. Biol. Med.* 41, 9 (2006).

[32] B. Chen, R. Yue, Y. Yang, H. Zeng, W. Chang, N. Gao, X. Yuan, W. Zhang, L. Shan. *Neurochem. Res.* (2014).

[33] W.X. Zheng, F. Wang, X.L. Cao, H.Y. Pan, X.Y. Liu, X.M. Hu, Y.Y. Sun. *Brain Inj.* 28, 2 (2014).

[34] X. Han, S. Zhu, B. Wang, L. Chen, R. Li, W. Yao, Z. Qu. *Neurochem. Int.* 64 (2014).

[35] X. Tian, L.P. Guo, X.L. Hu, J. Huang, Y.H. Fan, T.S. Ren, Q.C. Zhao. *Cell Mol. Neurobiol.* (2014).

[36] S. Seetapun, J. Yaoling, Y. Wang, Y.Z. Zhu. *Biosci. Rep.* 33, 4 (2013).

[37] L. Xie, L.F. Hu, X.Q. Teo, C.X. Tiong, V. Tazzari, A. Sparatore, P. Del Soldato, G.S. Dawe, J.S. Bian. *PLoS One.* 8, 4 (2013).

[38] S.F. Yang, Z.X. Yang, Q. Wu, A.S. Sun, X.N. Huang, J.S. Shi. *Acta Pharmcol. Sin.* 22, 12 (2001).

[39] Y. Pan, Y. Chen, Q. Li, X. Yu, J. Wang, J. Zheng. *Molecules.* 18, 2 (2013).

[40] G.F. Zeng, Z.Y. Zhang, L. Lu, D.Q. Xiao, S.H. Zong, J.M. He. *Rejuvenation Res.* 16, 2 (2013).

[41]  R. Yang, Q. Wang, L. Min, R. Sui, J. Li, X. Liu. *Neurol. Sci.* 34, 8 (2013).

[42]  S.C. Correia, R.X. Santos, M.S. Santos, G. Casadesus, J.C. Lamanna, G. Perry, M.A. Smith, P.I. Moreira. *Curr. Alzheimer Res.* 10, 4 (2013).

[43]  F.J. Miana-Mena, C. González-Mingot, P. Larrodé, M.J. Muñoz, S. Oliván, L. Fuentes-Broto, E. Martínez-Ballarín, R.J. Reiter, R. Osta, J.J. García. *J. Neurol.* 258, 5 (2011).

[44]  E.D. Hall, P.K. Andrus, J.A. Oostveen, T.J. Fleck, M.E. Gurney. *J. Neurosci. Res.* 53, 1 (1998).

[45]  O.P. Sunday, M.F. Adekunle, O.T. Temitope, A.A. Richard, A.A. Samuel, A.J. Olufunminyi, O.O. Elizabeth. *Int. J. Nutr. Pharmacol. Neurol. Dis.* 4, 3 (2014).

[46]  P. Bermejo, P. Gómez-Serranillos, J. Santos, E. Pastor, P. Gil, S. Martín-Aragón. *Gerontology.*43, 4 (1997).

[47]  G. Harish, C. Venkateshappa, A. Mahadevan, N. Pruthi, M.M. Srinivas Bharath, S.K. Shankar. *Neurochem. Int.* 59, 7 (2011).

[48]  P.S. Fitzmaurice, J. Tong, M. Yazdanpanah, P.P. Liu, K.S. Kalasinsky, S.J. Kish. *J. Pharmacol. Exp. Ther.* 319, 2 (2006).

[49]  N. Candan, N. Tuzmen. *Neurotoxicology.* 29, 4 (2008).

[50]  X.X. Ying, H.B. Li, Z.Y. Chu, Y.J. Zhai, A.J. Leng, X. Liu, C. Xin, W.J. Zhang, T.G. Kang. *Arch. Pharm. Res.* 31, 7 (2008).

[51]  S. Anil Kumar, S.A. Saif, P. Oothuman, M.I.A. Mustafa. *IMJM.* 10, 2 (2011).

[52]  K.P. Kuruvilla, M.S. Nandhu, J. Paul, C.S. Paulose. *J. Neurol. Sci.* 331, 1-2 (2013).

[53]  C.M. Lauderback, J.M. Hackett, J.N. Keller, S. Varadarajan, L. Szweda, M, Kindy, W.R. Markesbery, D.A. Butterfield. *Biochemistry.* 40, 8 (2001).

[54]  T. Ishrat, K. Parveen, M.N. Hoda, M.B. Khan, S. Yousuf, M.A. Ansari, S. Saleem, F. *Islam. Behav. Pharmacol.* 20, 7 (2009).

[55]  E.D. Hall, J.A. Oostveen, P.K. Andrus, D.K. Anderson, C.E. Thomas. *J. Neurosci. Methods.* 76, 2 (1997).

[56]  B. Uttara, A. V. Singh, P. Zamboni, R.T. Mahajan. *Curr. Neuropharmacol.* 7, 1, (2009).

[57]  M. M. Essa, R. K. Vijayan, G. Castellano-Gonzalez, M. A. Memon, N. Braidy, G. J. Guillemin. *Neurochem. Res.* 37, 9 (2012).

[58]  C. Guerra-Araiza, A. L. Álvarez-Mejía, S. Sánchez-Torres, E. Farfan-García, R. Mondragón-Lozano, R. Pinto-Almazán, H. Salgado-Ceballos. *Free Radic. Res.* 47, 6-7 (2013).

[59]  E. González-Burgos, M. P. Gómez-Serranillos.2014. In: M. Suzuki and S. Yamamoto (ed.) *Handbook on Reactive Oxygen Species.* Nova Biomedical, New York, chapter 7, pp. 245-265.

[60]  R. Sultana, M. Perluigi, D. Allan Butterfield. *Free Radic, Biol Med.* 62 (2013).

[61]  P. Devbhuti, A. Saha, C. Sengupta. *Acta Pol. Pharm.* 66, 4 (2009).

[62]  Y. Zhao, B. Zhao. *Front. Biosci.* (Elite Ed) 4 (2012).

[63]  Y. T. Choi, C. H. Jung, S. R. Lee, J. H. Bae, W. K. Baek, M. H. Suh, J. Park, C. W. Park, S. I. Suh. *Life Sci.* 70, 5 (2001).

[64]  J. Ma, W. Q. Yang, H. Zha, H. R. Yu. *Zhong Yao Cai.* 36, 2 (2013).

[65]  X. Tian, L. Zhang, J. Wang, J. Dai, S. Shen, L. Yang, P. Huang. *Behav. Brain Res.* 242 (2013).

[66]  M. R. Sepand, M. Soodi, H. Hajimehdipoor, M. Soleimani, E. Sahraei. *Iran J. Pharm. Res.* 12, 2 (2013).

[67]  S. Ouyang, L. S. Sun, S. L. Guo, X. Liu, J. P. Xu. Di Yi Jun Yi Da Xue Xue Bao 25, 2 (2005).

[68]  O. Cioanca, L. Hritcu, M. Mihasan, M. Hancianu. *Physiol. Behav.* 120 (2013).

[69]  Z. B. Liu, W. M. Niu, X. H. Yang, Y. Wang, W. G. Wang. *J. Tradit. Chin. Med.* 30, 4 (2010).

[70]  S. Li, X. P. Pu. *Biol. Pharm. Bull.* 34, 8 (2011).

[71]  M. T. Mansouri, Y. Farbood, M. J. Sameri, A. Sarkaki, B. Naghizadeh, M. Rafeirad. *Food Chem.* 138, 2-3 (2013).

[72]  S. F. Yang, Q. Wu, A. S. Sun, X. N. Huang, J. S. Shi. *Acta Pharmacol. Sin.* 22, 12 (2001).

[73]  L. Hritcu, H. S. Foyet, M. Stefan, M. Mihasan, A. E. Asongalem, P. Kamtchouing. *J. Ethnopharmacol.* 137, 1 (2011).

[74]  S. K. Yadav, J. Prakash, S. Chouhan, S. P. Singh. *Neurochem. Int.* 62, 8 (2013).

[75]  P. Kumar, S. S. Padi, P. S. Naidu, S. Kumar. Methods Find. *Exp. Clin. Pharmacol.* 29, 1 (2007).

[76]  H. M. Mahdy, M. G. Tadros, M. R. Mohamed, A. M. Karim, A. E. Khalifa. *Neurochem. Int.* 59, 6 (2011).

[77]  J. Zhang, J. Q. Yang, B. C. He, Q. X. Zhou, H. R. Yu, Y. Tang, B. Z. Liu. *Saudi Med. J.* 30, 6 (2009).

[78]  H. Y. Zhang, X. C. Tang. *Neurosci. Lett.* 292, 1 (2000).

[79]  B. Naghizadeh, M. T. Mansouri, B. Ghorbanzadeh, Y. Farbood, A. Sarkaki. *Phytomedicine* 20, 6 (2013).

[80] X. Lu, H. Xu, B. Sun, Z. Zhu, D. Zheng, X. Li. *Mol. Pharm.* 10, 5 (2013).

[81] K. Sun, S. Cao, L. Pei, A. Matsuura, L. Xiang, J. Qi. *Int. J. Mol. Sci.* 14, 3 (2013).

[82] Y. Sun, Y. Lin, X. Cao, L. Xiang, J. Qi. *Int. J. Mol. Sci.* 15, 12 (2014).

[83] W. Hu, G. Wang, P. Li, Y. Wang, C. L. Si, J. He, W. Long, Y. Bai, Z. Feng, X. Wang. *Chem. Biol. Interact.* 224C (2014).

[84] W. Fu, Y. Lei, J. L. Chen, C. M. Xiong, D. N. Zhou, G. H. Wu, J. Chen, Y. L. Cai, J. L. *Neurobiol. Learn. Mem.* 94, 3 (2010).

[85] N. Li, B. Liu, D. E. Dluzen, Y. Jin. *J. Ethnopharmacol.* 111, 3 (2007).

[86] J. H. Jeong, H. R. Jeong, Y. N. Jo, H. J. Kim, J. H. Shin, H. J. Heo HJ. *BMC Complement. Altern Med.* 13 (2013).

[87] M. Hancianu, O. Cioanca, M. Mihasan, L. Hritcu. *Phytomedicine* 20, 5 (2013)

[88] Y. Z. Shang, B. W. Qin, J. J. Cheng, H. Miao. *Phytother. Res.* 20, 1 (2006).

[89] G. F. Zeng, Z. Y. Zhang, L. Lu, D. Q. Xiao, S. H. Zong, J. M. He. *Rejuvenation Res.* 16, 2 (2013).

[90] E. González-Burgos, M. E. Carretero, M. P. Gómez-Serranillos. *Phytochemistry* 93 (2013).

[91] S. Z. Zhong, Q. H. Ge, R. Qu, Q. Li, S. P. Ma. *J. Neurol.* Sci. 277, 1-2 (2009).

[92] L. M. Chen, X. M. Zhou, Y. L. Cao, W. X. Hu. *J. Asian Nat. Prod. Res.* 10, 5-6 (2008).

[93] K. A. Koo, M. K. Lee, S. H. Kim, E. J. Jeong, S. Y. Kim, T. H. Oh, Y. C. Kim. *Br. J. Pharmacol.* 150, 1 (2007).

[94] S. Y. Li, Y. H. Jia, W. G. Sun, Y. Tang, G. S. An, J. H. Ni, H. T. Jia. Free Radic. Biol. Med. 48, 4 (2010).

[95] N. Limpeanchob, S. Jaipan, S. Rattanakaruna, W. Phrompittayarat, K. Ingkaninan. *J. Ethnopharmacol.* 120, 1 (2008).

[96] M. H. Veerendra Kumar, Y. K. Gupta. *Clin. Exp. Pharmacol. Physiol.*30, 5-6 (2003).

[97] S. Kaur, R. Chhabra, B. Nehru. *Phytomedicine* 20, 2 (2013).

[98] C. Sutalangka, J. Wattanathorn, S. Muchimapura, W. Thukham-mee. *Oxid. Med. Cell. Longev.* 2013 (2013).

[99] J. Y. Kim, H. Y. Jeong, H. K. Lee, S. Kim, B. Y. Hwang, K. Bae, Y. H. Seong. *Phytomedicine.* 19, 2 (2012).

[100] G. Oboh, A. J. Akinyemi, A. O. Ademiluyi. *Exp. Toxicol. Pathol.* 64, 1-2 (2012).

In: Malondialdehyde (MDA)    ISBN: 978-1-63482-793-5
Editor: Jackson Campbell    © 2015 Nova Science Publishers, Inc.

*Chapter 2*

# IMPACT OF MALONDIALDEHYDE ON COGNITIVE DYSFUNCTION IN OBESITY

***Wasana Pratchayasakul***[1,2]***, Nipon Chattipakorn***[1,2]
***and Siriporn C. Chattipakorn***[1,3,*]

[1]Neurophysiology Unit, Cardiac Electrophysiology Research and Training
Center, Faculty of Medicine, Chiang Mai University,
Chiang Mai, Thailand
[2]Department of Physiology, Faculty of Medicine, Chiang Mai University,
Chiang Mai, Thailand
[3]Department of Oral Biology and Diagnostic Science, Faculty of Dentistry,
Chiang Mai University, Chiang Mai, Thailand

## ABSTRACT

The incidence and level of obesity has increased dramatically in many countries around the world. Obesity escalates the risk of increased oxidative stress, which is believed to be one of the major underlying causes of many pathological conditions including insulin resistance, type II diabetes mellitus and cognitive decline. Malondialdehyde acts as a biomarker for oxidative stress. There is a great deal of evidence to show

[*] Corresponding author : Siriporn C. Chattipakorn, Department of Oral Biology and Diagnostic Science, Faculty of Dentistry, Chiang Mai University, and Neurophysiology Unit, Cardiac Electrophysiology Research and Training Center, Faculty of Medicine, Chiang Mai University, Chiang Mai, Thailand 50200; Tel: 011-66-53-944-451; Fax: 011-66-53-222-844; E-mail address:scchattipakorn@gmail.com; siriporn.c@cmu.ac.th.

the impact of the involvement of malondialdehyde in many pathological conditions. This chapter focuses on the role of malondialdehyde in the development of insulin resistance in obesity. Recent research in obese models has shown that increased oxidative stress can also lead to brain insulin resistance in addition to peripheral insulin resistance. The impairment of brain insulin sensitivity has been strongly implicated in the development of cognitive decline. This chapter also comprehensively summarizes the pathophysiological roles of malondialdehyde and its impact on cognitive function in obesity.

# 1. INTRODUCTION

The increase in prevalence and the severity of obesity has occurred too quickly to be attributed only to genetic factors. Obesity escalates the risk of developing insulin resistance, which is believed to play a major role in the development of certain pathological conditions such as diabetes and metabolic syndrome. Furthermore, previous studies found that weight gain following overfeeding significantly induced insulin resistance [1], whereas weight loss by caloric restriction reversed insulin resistance [2, 3]. Insulin resistance is a pathological state, in which target cells are resistant to the biological activities of insulin on glucose and lipid metabolism. Many insulin resistant individuals maintain near normal levels of plasma glucose by compensatory hyperinsulinemia [4, 5]. Although the common characteristics of insulin resistance are able to maintain the level of euglycemia, the combination of insulin resistance with hyperinsulinemia significantly increases the risk of developing some degree of glucose intolerance, high plasma triacylglycerol concentrations, low high-density lipoprotein (HDL-C) concentrations and essential hypertension. These abnormal conditions predispose individuals to be insulin resistant and increase their likelihood of developing type II diabetes [6-9].

Various pieces of evidence have shown that impaired insulin sensitivity is caused by several mechanisms, such as: lipotoxicity; adipocyte derived hormones; adipocyte cytokines and oxidative stress [10-12]. Among these mechanisms, oxidative stress, especially malondialdehyde, has been extensively studied for its involvement in the development of insulin resistance in obesity, which in turn can lead to type II diabetes. It has been demonstrated in obese models that increased oxidative stress can induce not only peripheral insulin resistance, but also brain insulin resistance [13-15]. The impairment of brain insulin sensitivity and the development of brain insulin resistance have been shown to contribute strongly to the development of cognitive decline. In this chapter, the pathophysiological roles of malondialdehyde, and its impact on cognitive function in obesity will be comprehensively summarized and discussed.

## 2. OBESITY, CIRCULATING MALONDIALDEHYDE LEVELS AND INSULIN RESISTANCE: POSSIBLE MECHANISMS AND CONSEQUENCES

The roles of oxidative stress in insulin resistance have been extensively investigated. An increase in the production of reactive oxygen species (ROS) has been shown to be associated with the mechanism causing obesity-induced insulin resistance. Several cellular sources for increased ROS generation in obesity-induced insulin resistance have been proposed. First, the increased circulating levels of pro-inflammatory cytokines which are shown to lead to the activation of the NADPH oxidases system and NOS-generating system [16, 17]. Second, the elevated circulating insulin levels which result in the activation of the NADPH oxidase system [18]. The dysregulation of carbohydrate and lipid metabolism has also been shown to cause an increased mitochondrial oxidation [19]. The activation of the NADPH oxidase system, NOS-generating system and increased mitochondrial oxidation have been demonstrated to lead to ROS generation [19]. An excess ROS generation can damage the membrane bilayers and cause lipid peroxidation. Lipid peroxidation is a free radical oxidation of polyunsaturated fatty acids (PUFA) in the fatty acid constituent of the membrane, a process which leads to the generation of reactive aldehydes, such as malondialdehyde (MDA), 4-hydroxynonenal (HNE) and acrolein [20]. These reactive aldehydes can form adducts with DNA and proteins, altering their function, and subsequently leading to the development of various diseases [21, 22].

MDA is an end-product generated by the decomposition of polyunsaturated fatty acids (PUFAs) through an enzymatic process via the biosynthesis of thromboxane A2 (TXA2) and 12-l-hydroxy-5,8,10-heptadecatrienoic acid (HHT); and a non-enzymatic process via bicyclic endoperoxides produced during lipid peroxidation [23]. Increased MDA levels (the biomarker of oxidative stress) have been shown to be associated with insulin resistance in several models [24-26]. In clinical studies of patients with impaired insulin sensitivity, [27] obesity [25], hyperlipidemia [28], metabolic syndrome (MetS) [26, 29, 30] or polycystic ovary syndrome (PCOS) [24, 31-33], all displayed characteristics of insulin resistance. Subjects with these pathological conditions showed increased MDA levels when compared with control subjects. Interestingly, the higher levels of MDA were associated with the severity of metabolic syndrome or polycystic ovary syndrome [26, 30, 32]. In addition, the study in PCOS patients found that the level of MDA showed a positive correlation with age, body mass index (BMI), waist-to-hip ratio, systolic and diastolic blood pressures, area under the curve (AUCs) for glucose, insulin level, insulin sensitivity index (ISI), triglyceride level [33] and the homeostasis model assessment-estimated insulin resistance (HOMA-IR) [24]. Moreover, MDA levels were also increased in obese children when compared to MDA in lean children [34-37]. This increase in MDA level was associated with increased levels of HOMA-IR [34] and metabolic risk factors [35]. Consistent with clinical studies, studies in animals found increased MDA levels in models of insulin resistance such as high fat diet-fed rats [38-45], high-carbohydrate and high-fat diet-fed rats [46, 47], Zucker fatty rats [48], ovariectomized rats [49], high fat diet-fed mice [50-52], obesogenic diet-fed swine [53] and modified atherogenic diet-fed swine [54]. In addition, this increased MDA level showed a positive correlation with body weight gain, serum leptin and HOMA index [44].

**Table 1. Evidence of increased levels of malondialdehyde (MDA) in subjects with insulin resistance (Clinical studies)**

| Subject | Tissue sample | Major findings | References |
|---|---|---|---|
| Human (insulin-resistant vs insulin-sensitive subjects) | Intestinal MDA | - Increased MDA level in insulin-resistant subjects, compared to insulin-sensitive subjects | Veilleux et al., 2014[24] |
| Human (patients with hyperlipidemia and normolipidemic controls) | Plasma MDA | - Increased MDA level in hyperlipidemia patients, compared to normolipidemic control<br>- MDA was inversely correlated with somatostatin levels | Yang et al., 2011[25] |
| Human (overweight/obese vs normal-weight controls in young and adult subjects) | Plasma MDA | - Increased MDA level only in overweight/obese with young MetS subjects | Mangge et al., 2013[22] |

| Subject | Tissue sample | Major findings | References |
|---|---|---|---|
| Human (MetS vs non-MetS subjects) | Serum MDA | - MetS was associated with a higher MDA level | Samaras et al., 2012[23] |
| Human (MetS vs non-MetS subjects) | Plasma MDA | - Increased MDA level in MetS subjects, compared to non-MetS subjects | Palomo et al., 2011[26] |
| Human (obese with MetS vs obese without MetS vs healthy controls) | Plasma MDA | - Increased MDA level in the obese subjects, especially in those exhibiting metabolic syndrome, compared to the healthy controls | Skalicky et al., 2008[27] |
| Human (polycystic ovary syndrome (PCOS) vs control women) | Serum MDA | - Increased MDA level in PCOS subjects, compared to control subjects | Liu et al., 2013[28] |
| Human (polycystic ovary syndrome (PCOS) vs control women) | Serum MDA | - Increased MDA level in PCOS subjects, compared to control subjects<br>- Increased MDA level was a risk factor for the development of PCOS | Fan et al., 2012[29] |
| Human (obese children vs normal weight children) | Plasma MDA | - Increased MDA level in obese children, compared to normal weight children | Codoñer-Franch et al., 2014[34] |
| Human (obese children with high values of HOMA-IR vs without high values of HOMA-IR) | Plasma MDA | - Higher MDA level in the group of children with greater HOMA-IR<br>- MDA level was positively correlated with HOMA-IR. | Codoñer-Franch et al., 2012[31] |
| Human (obese vs healthy children) | Plasma MDA | - Increased MDA level in obese children, compared to healthy children | Codoñer-Franch et al., 2012[33] |
| Human (obese children evaluated with respect to metabolic risk factors vs normal weight children) | Plasma MDA | - Increased MDA level in obese children, compared to normal weight children<br>- MDA level was associated with plasma nitrite | Codoñer-Franch et al., 2011[32] |

**Table 1. (Continued)**

| Subject | Tissue sample | Major findings | References |
|---|---|---|---|
| Human (non-obese women with polycystic ovary syndrome (PCOS) vs control women) | Plasma MDA | - MDA trend to increase in PCOS women, compared to controls<br>- MDA level positively correlated with HOMA-IR | Macut et al., 2011[21] |
| Human (polycystic ovary syndrome (PCOS) vs control women) | Erythrocyte MDA | - MDA trend to increase in PCOS women, compared to controls<br>- MDA levels positively correlated with age, BMI, waist-to-hip ratio, systolic and diastolic blood pressures, AUCs for glucose and insulin, ISI and triglyceride | Sabuncu et al., 2001[30] |
| Human (polycystic ovary syndrome (PCOS) vs control women) | Serum MDA | No significant differences were detected in MDA levels between groups | Karadeniz et al., 2011[53] |
| Human (non-obese vs frail lean vs frail obese elderly subjects) | Plasma MDA | No significant differences were detected in MDA level between groups | Goulet et al., 2009 [52] |

Although previous studies demonstrated that MDA levels were increased in insulin resistant models, some reports showed inconsistent findings. It has been shown that the MDA levels did not change in frail obese elderly subjects [55], PCOS patients [56] or high sucrose-fed rats [57], all of which are used as models of insulin resistance. Interestingly, the insulin levels were not elevated in these subjects when compared to the control subjects [55-57]. It is possible that the lack of change in the MDA level in these studies were due to the absence of hyperinsulinemia. The summarized information regarding reports on the MDA level in insulin-resistant conditions is shown in tables 1 and 2.

Despite the growing evidence that increased MDA level was related to insulin resistance, there are only a few reports that demonstrated the possible mechanism of MDA induced insulin resistance in obesity. Wang and colleagues demonstrated that moderate or high MDA levels promoted islet glucose-stimulated insulin secretion (GSIS) mainly through the Wnt signaling, elevating the ATP/ADP ratio and cytosolic $Ca^{2+}$ levels, and affecting the glucose transporter 2 (GLUT2), all of which resulted in hyperinsulinemia [58]. In addition, MDA was found to stimulate fatty acid synthesis and inhibit fatty acid beta-oxidation, via the inhibition of mitochondrial aconitase, leading to an accumulation of fat in adipose tissues [59]. These proposed mechanisms of obese-induced insulin resistance via increased MDA level are illustrated in figure 1.

**Table 2. Evidence of increased levels of malondialdehyde (MDA) in animal models with insulin resistance**

| Models | Tissue sample | Major findings | References |
|---|---|---|---|
| Rat (high fat diet vs normal diet –fed rats) | Plasma and brain MDA | - Increased MDA level in HFD-fed rats, compared to ND-fed rats | Pintana et al., 2013[37] |
| Rat (high fat diet vs normal diet –fed rats) | Serum MDA | - Increased MDA level in HFD-fed rats, compared to ND-fed rats | De Sibio et al., 2013[35] |
| Rat (high fat diet vs normal diet –fed rats) | Cadiac and liver MDA | - Increased MDA level in HFD-fed rats, compared to ND-fed rats | Lin et al., 2013[36] |
| Rat (high fat diet vs normal diet –fed rats) | Plasma and brain MDA | - Increased MDA level in HFD-fed rats, compared to ND-fed rats | Pintana et al., 2012[39] |
| Rat (high fat diet vs normal diet –fed rats) | Plasma and cardiac MDA | - Increased MDA level in HFD-fed rats, compared to ND-fed rats | Apaijai et al., 2012[38] |
| Rat (high fat diet vs normal diet –fed rats) | Liver MDA | - Increased MDA level in HFD-fed rats, compared to ND-fed rats <br> - MDA level was associated with apelin mRNA expression in white adipose tissue (WAT) | García-Díaz et al., 2007[40] |
| Rat (high fat diet vs normal diet –fed rats) | Liver MDA | - Increased MDA level in HFD-fed rats, compared to ND-fed rats <br> - MDA level correlated positively with body weight gain, serum leptin, and HOMA index | Milagro et al., 2006[41] |
| Rat (high-fat emulsion fed rats vs normal control rats) | Liver MDA | - Increased MDA level in HFD-fed rats, compared to ND-fed rats | Zou et al., 2006[42] |

**Table 2. (Continued)**

| Models | Tissue sample | Major findings | References |
|---|---|---|---|
| Rat (high-carbohydrate, high-fat diet vs normal diet -fed rats | Plasma MDA | - Increased MDA level in high-carbohydrate, high-fat diet-fed rats, compared to ND-fed rats | Panchal et al., 2011[43] |
| Rat (high-carbohydrate, high fat diet vs normal diet – fed rats) | Plasma MDA | - Increased MDA level in high-carbohydrate, high-fat diet-fed rats, compared to ND-fed rats | Poudyal et al., 2010[44] |
| Rat (Zucker fatty rats vs age-matched lean controls) | Plasma MDA | - Increased MDA level in zucker fatty rats, compared to age-matched lean rats | Mima et al., 2012[45] |
| Rat (Ovariectomized or sham-operated rats) | Serum and brain MDA | - Increased MDA level in OVX rats, compared to Sham rats<br>- High fat diet feeding accelerated an increase in MDA level in OVX rats | Pratchayasakul et al., 2014[46] |
| Mice (high fat diet vs normal diet –fed mice) | Serum and hepatic MDA | - Increased MDA level in HFD-fed rats, compared to ND-fed rats | Meng et al., 2011[49] |
| Mice (high fat diet vs normal diet –fed mice) | Plasma MDA | - Increased MDA level in HFD-fed rats, compared to ND-fed rats | de Oliveira et al., 2010[47] |
| Mice (high fat diet vs normal diet –fed mice) | Serum MDA | - Increased MDA level in HFD-fed rats, compared to ND-fed rats | Hagiwara et al., 2009[48] |
| Swine (obesogenic diets(fructose, atherogenic, and modified atherogenic ) vs normal diet –fed swine) | Serum and pancreatic MDA | - Increased MDA level in obesogenic diet -fed swine, compared to ND-fed swine | Fullenkamp et al., 2011[50] |
| Swine modified atherogenic (M-Ath) diet-fed swine vs normal diet –fed swine | Hepatic MDA | - Increased MDA level in M-Ath diet -fed swine, compared to ND-fed swine | Lee et al., 2009[51] |
| Rat (high sucrose diet vs normal diet –fed rats) | Liver MDA | - No significant differences were detected in MDA level | Lomba et al., 2009[54] |

*Abbreviation*: M-Ath diet, Diet with different sources of fat and higher protein, but lower choline content.

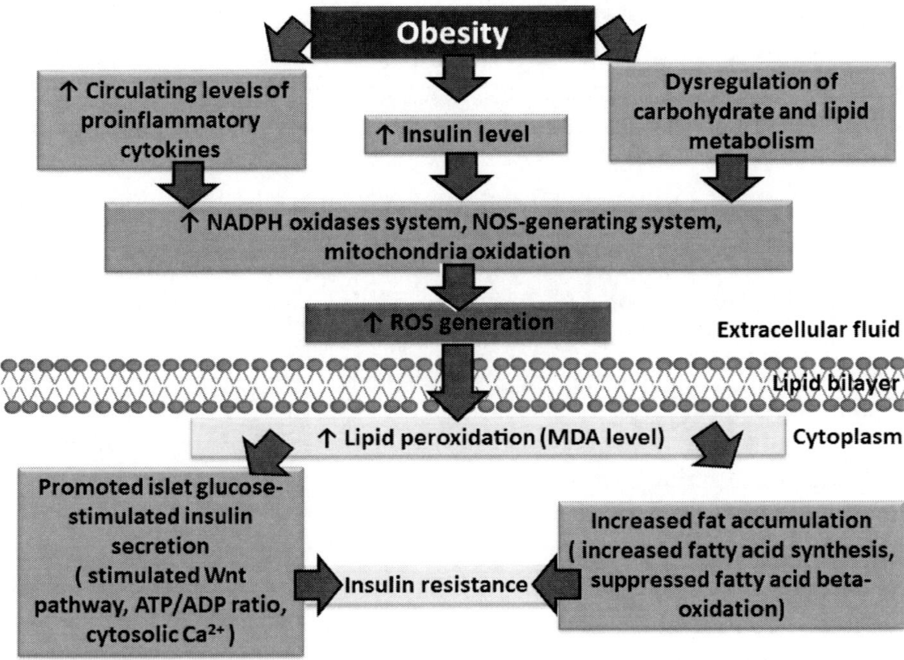


Figure 1. Proposed mechanism of obese induced peripheral insulin resistance via increased MDA level.

## 3. PERIPHERAL AND BRAIN INSULIN RESISTANCE, AND COGNITIVE DECLINE

The expression of insulin receptors are found in several areas of the brain, including the hippocampus and cerebral cortex, which have been shown to play important roles in learning and memory [60-62]. The beneficial effects of insulin on cognitive function have been demonstrated in both animal models and clinical studies [63-65]. In animal studies, an intra-cerebrovascular insulin injection improved learning and memory as demonstrated by a passive avoidance task [65] and Morris water-maze test [64, 66, 67]. In humans, intravenous infusion of insulin had positive effects on hippocampus-dependent types of declarative memory by enhancing word recall in a vigilance task [68]. In addition, intranasal insulin administration, which directly delivered insulin into the cerebrospinal fluid compartment, significantly improved word recall tasks and mood, including a reduction in anger and enhancement in self-confidence in a healthy subject [69-71]. Moreover, Benedict and colleagues found that the positive effects of insulin on cognitive functioning were more potent in women than in men, suggesting that neuronal insulin signaling may have gender-dependent property differences in controlling memory function [72].

Malfunctions in insulin signaling in the peripheral and neuronal tissues have been implicated in Alzheimer's disease, diabetes and aging [73-78]. In Alzheimer's disease, the reduction in cerebral insulin levels appears to be accompanied by functional disturbances of the insulin receptors (insulin-resistant brain state) [76]. In addition, the cross-sectional epidemiological studies in elderly humans have shown a longitudinal association between hyperinsulinemia or type II diabetes and Alzheimer's dementia [79-81]. In elderly non-demented and non-diabetic individuals, chronic hyperinsulinemia shows a correlation with cognitive decline [82]. Similarly, Young and colleagues demonstrated that middle-aged subjects, who did not have dementia, type II diabetes or stroke but with hyperinsulinemia, had a greater decline in performance in cognitive functioning tests over a 6-year period, compared with subjects without hyperinsulinemia [83]. All of these reports strongly indicate a negative association between hyperinsulinemia and cognitive function.

One of the critical functions of insulin on cognitive function is its role in an increase synaptic plasticity. Synaptic plasticity is the main mechanism for learning and memory processes [84, 85]. Two opposite forms of activity-dependent synaptic plasticity have been identified: long-term potentiation (LTP) and long-term depression (LTD) [84, 85]. LTP and LTD are initiated by many factors. A large rise in intracellular calcium levels leads to a stronger depolarization and brief high-frequency (100 Hz) stimulation, subsequently leading to a long lasting increase in the strength of synaptic transmission or LTP. On the other hand, a small rise in intracellular calcium levels leads to a lower level of post-synaptic depolarization and prolonged low-frequency (1Hz) stimulation, resulting in a persistent reduction in synaptic strength or LTD [84, 86, 87]. Both LTP and LTD are the primary experimental models for investigating the synaptic basis of learning and memory processes [85]. Several studies showed that a brief bath application of insulin can induce a long-lasting depression of excitatory synaptic transmission in the hippocampal CA3-CA1 Schaffer collateral pathway (insulin-induced LTD) [88-91].

In the insulin resistance model of fructose-fed hamsters, it has been shown that hippocampal synaptic plasticity was impaired after fructose feeding [92]. Furthermore, in both insulin resistant and type II diabetes models induced by long-term high-fat diet consumption, it has been shown that consumption of a high-fat diet reduced LTP in the CA1 of the hippocampus, and impaired cognitive functioning in rats [93]. In addition, long term consumption of high-fat diet also caused peripheral insulin resistance and impeded cognitive performance [94-98]. These findings suggest that the development of insulin resistance can mediate the cognitive deficit associated with a high-fat diet. Recently, our studies have shown that consumption of a high-fat diet for 12 weeks in male and female rats could cause not only peripheral insulin resistance but also brain insulin resistance, indicated by the impairment of brain insulin receptor function (insulin-induced LTD) [13-15, 99]. The reduced ability of insulin-induced LTD was caused by the decrease in the phosphorylation of the insulin receptors (IR), insulin receptor substrate (IRS-1) and serine/threonine protein kinase/protein kinase B (Akt/PKB), in brain slices of these obese-insulin resistant rats [13-15, 99].

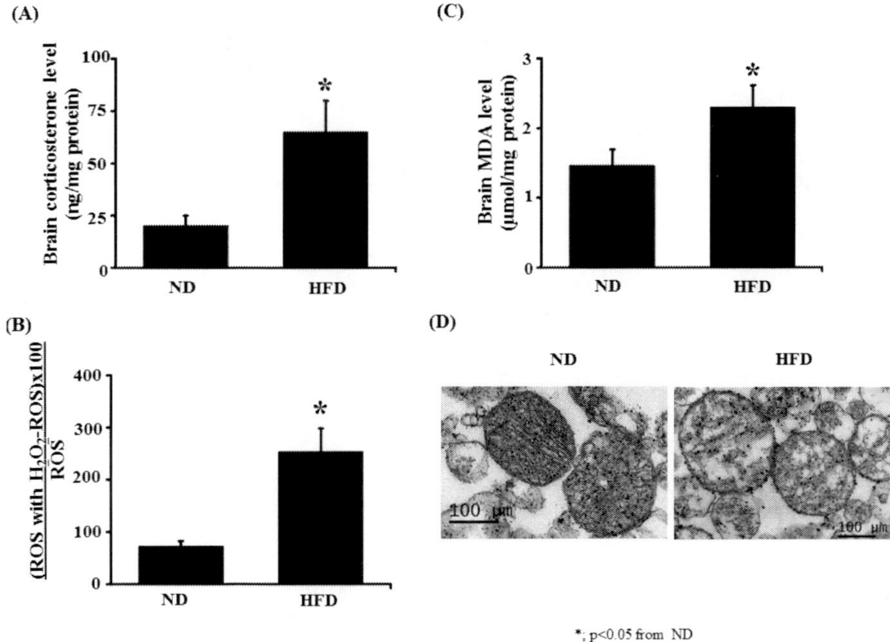


Figure 2. Long term high-fat diet consumption increased brain oxidative stress in rats. (A) Brain corticosterone level, (B) Brain mitochondrial ROS production, (C) Brain MDA level and (D) Brain mitochondrial morphology from the ND and HFD groups. *, p<0.05 compared with ND: ND= normal diet-fed rats, HFD= high fat diet-fed rats.

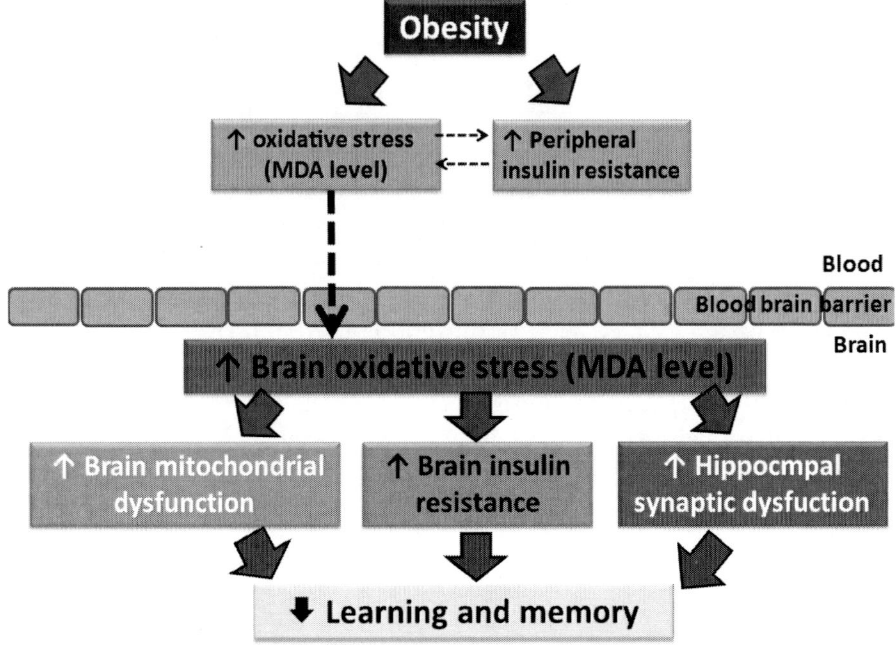


Figure 3. Proposed mechanism of obese induced insulin resistance impaired cognitive function via increased MDA level.

## 4. THE ASSOCIATION OF MALONDIALDEHYDE LEVELS AND COGNITIVE DECLINE UNDER OBESE-INSULIN RESISTANT CONDITIONS

Although MDA has been found in the brain of several neurodegenerative diseases, including Alzheimer's disease, Parkinson's disease and amyotrophic lateral sclerosis [23, 100], its association with cognition in the obese-insulin resistant models was previously unclear. Recently, we demonstrated that obese-insulin resistant rats had increased brain oxidative stress as indicated by increased brain corticosterone levels [14], brain MDA level [40, 42], brain mitochondrial reactive oxygen species (ROS) production and also brain mitochondrial swelling [13, 15, 40, 42] (Figure 2). The increased brain oxidative stress and impaired brain mitochondrial function have been proposed as the underlying mechanisms for brain insulin resistance, as indicated by the impaired ability of insulin-induced LTD and impaired brain insulin signaling [13-15, 40, 49]. These mechanisms are proposed to be mainly responsible for impaired cognition observed in obese-insulin resistant models as indicated by the increased time to reach the platform and decreased time of stay in the target quadrant in the Morris Water Maze test [13, 40, 42]. These mechanisms are illustrated in Figure 3.

## CONCLUSION

In obese-insulin resistant subjects, increased peripheral insulin resistance and oxidative stress could lead to brain oxidative stress, impaired brain insulin receptor signaling (or brain insulin resistance), brain mitochondrial dysfunction and reduced brain synaptic plasticity. These impairments ultimately lead to cognitive decline in obese-insulin resistant subjects.

## ACKNOWLEDGEMENTS

This work was supported by Thailand Research Fund TRF-BRG5780016 (SC), TRF-TRG5680018 (WP), Faculty of Medicine Chiang Mai University Endowment Fund (NC), National Research Council of Thailand (SC), Chiang Mai University Center of Excellence Award (NC) and the NSTDA Research Chair Grant from the National Science and Technology Development Agency (NC). The authors would like to thank Ms. Aimee Cole for her editorial assistance.

## REFERENCES

[1]     Ravussin, E., Smith, S. R. (2002) Increased fat intake, impaired fat oxidation, and failure of fat cell proliferation result in ectopic fat storage, insulin resistance, and type 2 diabetes mellitus. *Ann N Y Acad Sci, 967*, 363-378.

[2]     Albu, J., Reed, G. (1995) Weight cycling: more questions than answers. *Endocr Pract, 1*, 346-352.

[3]     Henry, R. R., Wallace, P., Olefsky, J. M. (1986) Effects of weight loss on mechanisms of hyperglycemia in obese non-insulin-dependent diabetes mellitus. *Diabetes, 35*, 990-998.

[4]     DeFronzo, R. A. (1992) Pathogenesis of type 2 (non-insulin dependent) diabetes mellitus: a balanced overview. *Diabetologia, 35*, 389-397.

[5]     DeFronzo, R. A., Bonadonna, R. C., Ferrannini, E. (1992) Pathogenesis of NIDDM. A balanced overview. *Diabetes Care, 15*, 318-368.

[6]     Areosa, S. A., Grimley, E. V. (2002) Effect of the treatment of Type II diabetes mellitus on the development of cognitive impairment and dementia. *Cochrane Database Syst Rev*, CD003804.

[7]     Cole, G. M., Lim, G. P., Yang, F., et al. (2005) Prevention of Alzheimer's disease: Omega-3 fatty acid and phenolic anti-oxidant interventions. *Neurobiol Aging, 26*, 133-136.

[8]     Conquer, J. A., Tierney, M. C., Zecevic, J., Bettger, W. J., Fisher, R. H. (2000) Fatty acid analysis of blood plasma of patients with Alzheimer's disease, other types of dementia, and cognitive impairment. *Lipids, 35*, 1305-1312.

[9]     Elias, M. F., Elias, P. K., Sullivan, L. M., Wolf, P. A., D'Agostino, R. B. (2005) Obesity, diabetes and cognitive deficit: The Framingham Heart Study. *Neurobiol Aging, 26 Suppl 1*, 11-16.

[10]    Boden, G. (1997) Role of fatty acids in the pathogenesis of insulin resistance and NIDDM. *Diabetes, 46*, 3-10.

[11]    Fujimoto, W. Y. (2000) The importance of insulin resistance in the pathogenesis of type 2 diabetes mellitus. *Am J Med, 108*, 9S-14S.

[12]    Ye, J. (2013) Mechanisms of insulin resistance in obesity. *Front Med, 7*, 14-24.

[13]    Pipatpiboon, N., Pintana, H., Pratchayasakul, W., Chattipakorn, N., Chattipakorn, S. C. (2013) DPP4-inhibitor improves neuronal insulin receptor function, brain mitochondrial function and cognitive function in rats with insulin resistance induced by high-fat diet consumption. *Eur J Neurosci, 37*, 839-849.

[14]    Pratchayasakul, W., Kerdphoo, S., Petsophonsakul, P., Pongchaidecha, A., Chattipakorn, N., Chattipakorn, S. C. (2011) Effects of high-fat diet on insulin receptor function in rat hippocampus and the level of neuronal corticosterone. *Life Sci, 88*, 619-627.

[15]    Pipatpiboon, N., Pratchayasakul, W., Chattipakorn, N., Chattipakorn, S. C. (2012) PPARgamma agonist improves neuronal insulin receptor function in hippocampus and brain mitochondria function in rats with insulin resistance induced by long term high-fat diets. *Endocrinology, 153*, 329-338.

[16]   Omori, K., Ohira, T., Uchida, Y., et al. (2008) Priming of neutrophil oxidative burst in diabetes requires preassembly of the NADPH oxidase. *J Leukoc Biol, 84,* 292-301.

[17]   Karima, M., Kantarci, A., Ohira, T., et al. (2005) Enhanced superoxide release and elevated protein kinase C activity in neutrophils from diabetic patients: association with periodontitis. *J Leukoc Biol, 78,* 862-870.

[18]   Mahadev, K., Motoshima, H., Wu, X., et al. (2004) The NAD(P)H oxidase homolog Nox4 modulates insulin-stimulated generation of H2O2 and plays an integral role in insulin signal transduction. *Mol Cell Biol, 24,* 1844-1854.

[19]   Bashan, N., Kovsan, J., Kachko, I., Ovadia, H., Rudich, A. (2009) Positive and negative regulation of insulin signaling by reactive oxygen and nitrogen species. *Physiol Rev, 89,* 27-71.

[20]   Mylonas, C., Kouretas, D. (1999) Lipid peroxidation and tissue damage. *In Vivo, 13,* 295-309.

[21]   Romero, F. J., Bosch-Morell, F., Romero, M. J., et al. (1998) Lipid peroxidation products and antioxidants in human disease. *Environ Health Perspect, 106,* 1229-1234.

[22]   Sultana, R., Perluigi, M., Allan Butterfield, D. (2013) Lipid peroxidation triggers neurodegeneration: a redox proteomics view into the Alzheimer disease brain. *Free Radic Biol Med, 62,* 157-169.

[23]   Ayala, A., Munoz, M. F., Arguelles, S. (2014) Lipid peroxidation: production, metabolism, and signaling mechanisms of malondialdehyde and 4-hydroxy-2-nonenal. *Oxid Med Cell Longev, 2014,* 360438.

[24]   Macut, D., Simic, T., Lissounov, A., et al. (2011) Insulin resistance in non-obese women with polycystic ovary syndrome: relation to byproducts of oxidative stress. *Exp Clin Endocrinol Diabetes, 119,* 451-455.

[25]   Mangge, H., Zelzer, S., Puerstner, P., et al. (2013) Uric acid best predicts metabolically unhealthy obesity with increased cardiovascular risk in youth and adults. *Obesity, 21,* E71-77.

[26]   Samaras, K., Crawford, J., Baune, B. T., et al. (2012) The value of the metabolic syndrome concept in elderly adults: is it worth less than the sum of its parts? *J Am Geriatr Soc, 60,* 1734-1741.

[27]   Veilleux, A., Grenier, E., Marceau, P., Carpentier, A. C., Richard, D., Levy, E. (2014) Intestinal lipid handling: evidence and implication of insulin signaling abnormalities in human obese subjects. *Arterioscler Thromb Vasc Biol, 34,* 644-653.

[28]  Yang, R. L., Li, W., Yue, P., Shi, Y. H., Le, G. W. (2011) Relation of plasma somatostatin levels with malondialdehyde in hyperlipidemic patients. *Asia Pac J Clin Nutr, 20*, 220-224.

[29]  Palomo, I., Contreras, A., Alarcon, L. M., et al. (2011) Elevated concentration of asymmetric dimethylarginine (ADMA) in individuals with metabolic syndrome. *Nitric Oxide, 24*, 224-228.

[30]  Skalicky, J., Muzakova, V., Kandar, R., Meloun, M., Rousar, T., Palicka, V. (2008) Evaluation of oxidative stress and inflammation in obese adults with metabolic syndrome. *Clin Chem Lab Med, 46*, 499-505.

[31]  Liu, H. W., Zhang, F., Fan, P., Bai, H., Zhang, J. X., Wang, Y. (2013) Effects of apolipoprotein E genotypes on metabolic profile and oxidative stress in southwest Chinese women with polycystic ovary syndrome. *Eur J Obstet Gynecol Reprod Biol, 170*, 146-151.

[32]  Fan, P., Liu, H., Wang, Y., Zhang, F., Bai, H. (2012) Apolipoprotein E-containing HDL-associated platelet-activating factor acetylhydrolase activities and malondialdehyde concentrations in patients with PCOS. *Reprod Biomed Online, 24*, 197-205.

[33]  Sabuncu, T., Vural, H., Harma, M., Harma, M. (2001) Oxidative stress in polycystic ovary syndrome and its contribution to the risk of cardiovascular disease. *Clin Biochem, 34*, 407-413.

[34]  Codoner-Franch, P., Navarro-Ruiz, A., Fernandez-Ferri, M., Arilla-Codoner, A., Ballester-Asensio, E., Valls-Belles, V. (2012) A matter of fat: insulin resistance and oxidative stress. *Pediatr Diabetes, 13*, 392-399.

[35]  Codoner-Franch, P., Tavarez-Alonso, S., Murria-Estal, R., Megias-Vericat, J., Tortajada-Girbes, M., Alonso-Iglesias, E. (2011) Nitric oxide production is increased in severely obese children and related to markers of oxidative stress and inflammation. *Atherosclerosis, 215*, 475-480.

[36]  Codoner-Franch, P., Tavarez-Alonso, S., Murria-Estal, R., Tortajada-Girbes, M., Simo-Jorda, R., Alonso-Iglesias, E. (2012) Elevated advanced oxidation protein products (AOPPs) indicate metabolic risk in severely obese children. *Nutr Metab Cardiovasc Dis, 22*, 237-243.

[37]  Codoner-Franch, P., Tavarez-Alonso, S., Porcar-Almela, M., Navarro-Solera, M., Arilla-Codoner, A., Alonso-Iglesias, E. (2014) Plasma resistin levels are associated with homocysteine, endothelial activation, and nitrosative stress in obese youths. *Clin Biochem, 47*, 44-48.

[38] De Sibio, M. T., Luvizotto, R. A., Olimpio, R. M., et al. (2013) A comparative genotoxicity study of a supraphysiological dose of triiodothyronine (T(3)) in obese rats subjected to either calorie-restricted diet or hyperthyroidism. *PLoS One, 8*, e56913.

[39] Lin, M. C., Hsu, P. C., Yin, M. C. (2013) Protective effects of Houttuynia cordata aqueous extract in mice consuming a high saturated fat diet. *Food Funct, 4*, 322-327.

[40] Pintana, H., Apaijai, N., Chattipakorn, N., Chattipakorn, S. C. (2013) DPP-4 inhibitors improve cognition and brain mitochondrial function of insulin-resistant rats. *J Endocrinol, 218*, 1-11.

[41] Apaijai, N., Pintana, H., Chattipakorn, S. C., Chattipakorn, N. (2012) Cardioprotective effects of metformin and vildagliptin in adult rats with insulin resistance induced by a high-fat diet. *Endocrinology, 153*, 3878-3885.

[42] Pintana, H., Apaijai, N., Pratchayasakul, W., Chattipakorn, N., Chattipakorn, S. C. (2012) Effects of metformin on learning and memory behaviors and brain mitochondrial functions in high fat diet induced insulin resistant rats. *Life Sci, 91*, 409-414.

[43] Garcia-Diaz, D., Campion, J., Milagro, F. I., Martinez, J. A. (2007) Adiposity dependent apelin gene expression: relationships with oxidative and inflammation markers. *Mol Cell Biochem, 305*, 87-94.

[44] Milagro, F. I., Campion, J., Martinez, J. A. (2006) Weight gain induced by high-fat feeding involves increased liver oxidative stress. *Obesity (Silver Spring), 14*, 1118-1123.

[45] Zou, Y., Li, J., Lu, C., et al. (2006) High-fat emulsion-induced rat model of nonalcoholic steatohepatitis. *Life Sci, 79*, 1100-1107.

[46] Panchal, S. K., Poudyal, H., Iyer, A., et al. (2011) High-carbohydrate, high-fat diet-induced metabolic syndrome and cardiovascular remodeling in rats. *J Cardiovasc Pharmacol, 57*, 611-624.

[47] Poudyal, H., Campbell, F., Brown, L. (2010) Olive leaf extract attenuates cardiac, hepatic, and metabolic changes in high carbohydrate-, high fat-fed rats. *J Nutr, 140*, 946-953.

[48] Mima, A., Qi, W., Hiraoka-Yamomoto, J., et al. (2012) Retinal not systemic oxidative and inflammatory stress correlated with VEGF expression in rodent models of insulin resistance and diabetes. *Invest Ophthalmol Vis Sci, 53*, 8424-8432.

[49] Pratchayasakul, W., Chattipakorn, N., Chattipakorn, S. C. (2014) Estrogen restores brain insulin sensitivity in ovariectomized non-obese rats, but not in ovariectomized obese rats. *Metabolism, 63*, 851-859.

[50] de Oliveira, P. R., da Costa, C. A., de Bem, G. F., et al. (2010) Effects of an extract obtained from fruits of Euterpe oleracea Mart. in the components of metabolic syndrome induced in C57BL/6J mice fed a high-fat diet. *J Cardiovasc Pharmacol, 56*, 619-626.

[51] Hagiwara, S., Gohda, T., Tanimoto, M., et al. (2009) Effects of pyridoxamine (K-163) on glucose intolerance and obesity in high-fat diet C57BL/6J mice. *Metabolism, 58*, 934-945.

[52] Meng, R., Zhu, D. L., Bi, Y., Yang, D. H., Wang, Y. P. (2011) Anti-oxidative effect of apocynin on insulin resistance in high-fat diet mice. *Ann Clin Lab Sci, 41*, 236-243.

[53] Fullenkamp, A. M., Bell, L. N., Robbins, R. D., et al. (2011) Effect of different obesogenic diets on pancreatic histology in Ossabaw miniature swine. *Pancreas, 40*, 438-443.

[54] Lee, L., Alloosh, M., Saxena, R., et al. (2009) Nutritional model of steatohepatitis and metabolic syndrome in the Ossabaw miniature swine. *Hepatology, 50*, 56-67.

[55] Goulet, E. D., Hassaine, A., Dionne, I. J., et al. (2009) Frailty in the elderly is associated with insulin resistance of glucose metabolism in the postabsorptive state only in the presence of increased abdominal fat. *Exp Gerontol, 44*, 740-744.

[56] Karadeniz, M., Erdogan, M., Ayhan, Z., et al. (2011) Effect Of G2706A and G1051A polymorphisms of the ABCA1 gene on the lipid, oxidative stress and homocystein levels in Turkish patients with polycystic ovary syndrome. *Lipids Health Dis, 10*, 193.

[57] Lomba, A., Milagro, F. I., Garcia-Diaz, D. F., Campion, J., Marzo, F., Martinez, J. A. (2009) A high-sucrose isocaloric pair-fed model induces obesity and impairs NDUFB6 gene function in rat adipose tissue. *J Nutrigenet Nutrigenomics, 2*, 267-272.

[58] Wang, X., Lei, X. G., Wang, J. (2014) Malondialdehyde regulates glucose-stimulated insulin secretion in murine islets via TCF7L2-dependent Wnt signaling pathway. *Mol Cell Endocrinol, 382*, 8-16.

[59] Nicolas, J. P. & Christophe, O. S. (2012) Lipid Peroxidation by-Products and the Metabolic Syndrome. In: Angel, C, editor. Lipid peroxidation: InTech; 2012; 400-436.

[60] Marks, J. L., Porte, D., Jr., Stahl, W. L., Baskin, D. G. (1990) Localization of insulin receptor mRNA in rat brain by in situ hybridization. *Endocrinology, 127*, 3234-3236.

[61] Schwartz, M. W., Figlewicz, D. P., Baskin, D. G., Woods, S. C., Porte, D., Jr. (1992) Insulin in the brain: a hormonal regulator of energy balance. *Endocr Rev, 13*, 387-414.

[62] Unger, J. W., Betz, M. (1998) Insulin receptors and signal transduction proteins in the hypothalamo-hypophyseal system: a review on morphological findings and functional implications. *Histol Histopathol, 13*, 1215-1224.

[63] Benedict, C., Frey, W. H., 2nd, Schioth, H. B., Schultes, B., Born, J., Hallschmid, M. (2011) Intranasal insulin as a therapeutic option in the treatment of cognitive impairments. *Exp Gerontol, 46*, 112-115.

[64] Moosavi, M., Naghdi, N., Maghsoudi, N., Zahedi Asl, S. (2006) The effect of intrahippocampal insulin microinjection on spatial learning and memory. *Horm Behav, 50*, 748-752.

[65] Park, C. R., Seeley, R. J., Craft, S., Woods, S. C. (2000) Intracerebroventricular insulin enhances memory in a passive-avoidance task. *Physiol Behav, 68*, 509-514.

[66] Haj-ali, V., Mohaddes, G., Babri, S. H. (2009) Intracerebroventricular insulin improves spatial learning and memory in male Wistar rats. *Behav Neurosci, 123*, 1309-1314.

[67] Moosavi, M., Naghdi, N., Choopani, S. (2007) Intra CA1 insulin microinjection improves memory consolidation and retrieval. *Peptides, 28*, 1029-1034.

[68] Kern, W., Peters, A., Fruehwald-Schultes, B., Deininger, E., Born, J., Fehm, H. L. (2001) Improving influence of insulin on cognitive functions in humans. *Neuroendocrinology, 74*, 270-280.

[69] Benedict, C., Hallschmid, M., Hatke, A., et al. (2004) Intranasal insulin improves memory in humans. *Psychoneuroendocrinology, 29*, 1326-1334.

[70] Benedict, C., Hallschmid, M., Schmitz, K., et al. (2007) Intranasal insulin improves memory in humans: superiority of insulin aspart. *Neuropsychopharmacology, 32*, 239-243.

[71] Benedict, C., Hallschmid, M., Schultes, B., Born, J., Kern, W. (2007) Intranasal insulin to improve memory function in humans. *Neuroendocrinology, 86*, 136-142.

[72] Benedict, C., Kern, W., Schultes, B., Born, J., Hallschmid, M. (2008) Differential sensitivity of men and women to anorexigenic and memory-improving effects of intranasal insulin. *J Clin Endocrinol Metab, 93*, 1339-1344.

[73] Biessels, G. J., van der Heide, L. P., Kamal, A., Bleys, R. L., Gispen, W. H. (2002) Ageing and diabetes: implications for brain function. *Eur J Pharmacol, 441*, 1-14.

[74] Craft, S., Peskind, E., Schwartz, M. W., Schellenberg, G. D., Raskind, M., Porte, D., Jr. (1998) Cerebrospinal fluid and plasma insulin levels in Alzheimer's disease: relationship to severity of dementia and apolipoprotein E genotype. *Neurology, 50*, 164-168.

[75] Craft, S., Watson, G. S. (2004) Insulin and neurodegenerative disease: shared and specific mechanisms. *Lancet Neurol, 3*, 169-178.

[76] Frolich, L., Blum-Degen, D., Bernstein, H. G., et al. (1998) Brain insulin and insulin receptors in aging and sporadic Alzheimer's disease. *J Neural Transm, 105*, 423-438.

[77] Gispen, W. H., Biessels, G. J. (2000) Cognition and synaptic plasticity in diabetes mellitus. *Trends Neurosci, 23*, 542-549.

[78] Hoyer, S. (1998) Is sporadic Alzheimer disease the brain type of non-insulin dependent diabetes mellitus? A challenging hypothesis. *J Neural Transm, 105*, 415-422.

[79] Kalmijn, S., Feskens, E. J., Launer, L. J., Kromhout, D. (1997) Polyunsaturated fatty acids, antioxidants, and cognitive function in very old men. *Am J Epidemiol, 145*, 33-41.

[80] Morris, M. C., Evans, D. A., Bienias, J. L., Tangney, C. C., Wilson, R. S. (2004) Dietary fat intake and 6-year cognitive change in an older biracial community population. *Neurology, 62*, 1573-1579.

[81] Ortega, R. M., Requejo, A. M., Andres, P., et al. (1997) Dietary intake and cognitive function in a group of elderly people. *Am J Clin Nutr, 66*, 803-809.

[82] Kalmijn, S., Feskens, E. J., Launer, L. J., Stijnen, T., Kromhout, D. (1995) Glucose intolerance, hyperinsulinaemia and cognitive function in a general population of elderly men. *Diabetologia, 38*, 1096-1102.

[83] Young, S. E., Mainous, A. G., 3rd, Carnemolla, M. (2006) Hyperinsulinemia and cognitive decline in a middle-aged cohort. *Diabetes Care, 29*, 2688-2693.

[84] Dunwiddie, T., Lynch, G. (1978) Long-term potentiation and depression of synaptic responses in the rat hippocampus: localization and frequency dependency. *J Physiol, 276*, 353-367.

[85] Malenka, R. C., Nicoll, R. A. (1999) Long-term potentiation--a decade of progress? *Science, 285*, 1870-1874.

[86]  Artola, A., Brocher, S., Singer, W. (1990) Different voltage-dependent thresholds for inducing long-term depression and long-term potentiation in slices of rat visual cortex. *Nature, 347*, 69-72.

[87]  Ngezahayo, A., Schachner, M., Artola, A. (2000) Synaptic activity modulates the induction of bidirectional synaptic changes in adult mouse hippocampus. *J Neurosci, 20*, 2451-2458.

[88]  Huang, C. C., Lee, C. C., Hsu, K. S. (2004) An investigation into signal transduction mechanisms involved in insulin-induced long-term depression in the CA1 region of the hippocampus. *J Neurochem, 89*, 217-231.

[89]  Huang, C. C., You, J. L., Lee, C. C., Hsu, K. S. (2003) Insulin induces a novel form of postsynaptic mossy fiber long-term depression in the hippocampus. *Mol Cell Neurosci, 24*, 831-841.

[90]  Man, H. Y., Lin, J. W., Ju, W. H., et al. (2000) Regulation of AMPA receptor-mediated synaptic transmission by clathrin-dependent receptor internalization. *Neuron, 25*, 649-662.

[91]  van der Heide, L. P., Kamal, A., Artola, A., Gispen, W. H., Ramakers, G. M. (2005) Insulin modulates hippocampal activity-dependent synaptic plasticity in a N-methyl-d-aspartate receptor and phosphatidyl-inositol-3-kinase-dependent manner. *J Neurochem, 94*, 1158-1166.

[92]  Mielke, J. G., Taghibiglou, C., Liu, L., et al. (2005) A biochemical and functional characterization of diet-induced brain insulin resistance. *J Neurochem, 93*, 1568-1578.

[93]  Stranahan, A. M., Norman, E. D., Lee, K., et al. (2008) Diet-induced insulin resistance impairs hippocampal synaptic plasticity and cognition in middle-aged rats. *Hippocampus, 18*, 1085-1088.

[94]  Molteni, R., Barnard, R. J., Ying, Z., Roberts, C. K., Gomez-Pinilla, F. (2002) A high-fat, refined sugar diet reduces hippocampal brain-derived neurotrophic factor, neuronal plasticity, and learning. *Neuroscience, 112*, 803-814.

[95]  Solfrizzi, V., Panza, F., Capurso, A. (2003) The role of diet in cognitive decline. *J Neural Transm, 110*, 95-110.

[96]  Kalmijn, S., van Boxtel, M. P., Ocke, M., Verschuren, W. M., Kromhout, D., Launer, L. J. (2004) Dietary intake of fatty acids and fish in relation to cognitive performance at middle age. *Neurology, 62*, 275-280.

[97]  Greenwood, C. E., Winocur, G. (2005) High-fat diets, insulin resistance and declining cognitive function. *Neurobiol Aging, 26*, 42-45.

[98]    Winocur, G., Greenwood, C. E. (2005) Studies of the effects of high fat diets on cognitive function in a rat model. *Neurobiol Aging, 26,* 46-49.

[99]    Pratchayasakul, W., Chattipakorn, N., Chattipakorn, S. C. (2011) Effects of estrogen in preventing neuronal insulin resistance in hippocampus of obese rats are different between genders. *Life Sci, 89,* 702-707.

[100]  Dib, M., Garrel, C., Favier, A., Robin, V., Desnuelle, C. (2002) Can malondialdehyde be used as a biological marker of progression in neurodegenerative disease? *J Neurol, 249,* 367-374.

In: Malondialdehyde (MDA)  ISBN: 978-1-63482-793-5
Editor: Jackson Campbell  © 2015 Nova Science Publishers, Inc.

*Chapter 3*

# GLUTATHIONE AND LIPOIC ACID BENEFIT EFFECTS ON LIVER, KIDNEY, BRAIN AND PANCREATIC TISSUE FROM Cd-, Pb- AND Cu- PROVOKED LIPID PEROXIDATION MONITORING VIA MDA CONTENT AMONG WISTAR RATS

*Ružica S. Nikolić[1], Jasmina M. Jovanović[2], Gordana M. Kocić[3] and Nenad S. Krstić[1]*
[1]Department of Chemistry, Faculty of Sciences and Mathematics, University of Niš, Niš, Serbia
[2]Higher Education School for Medical Professionals, Ćuprija, Serbia
[3]Faculty of Medicine, University of Niš, Niš, Serbia

## ABSTRACT

In this study was determined a content of MDA, as a marker of lipid peroxidation, in the internal organs (liver, kidneys, brain and pancreas) of heavy metal intoxicated Wistar rats. The potential protective effects of O- and S- donor supplements (glutathione and lipoic acid) on lipid peroxidation effects were also examined.

The content of malondialdehyde (MDA) was increased several times in the investigated tissues of exposed rats. The toxic effects of investigated metals being more pronounced in liver, respectively by metal: Pb > Cd > Cu. In other organs the toxic effect of metals, according of the level of MDA is respectively: Cd > Pb > Cu.

The treatment of intoxicated animals with glutathione and lipoic acid drastically suppressed lipid peroxidation. These supplements can form via their O- and S- donor atoms stable associations with the ions of these metals, thus blocking the heavy metals and reducing their toxic effects in the following order Cd > Pb > Cu.

FTIR analysis on the example of associates between copper and oxide and reduce form of lipoic acid, shows that reduced form react with S- donor atoms of free HS- groups, therefore oxide form react via O- donor atoms of carboxylic gropu.

The intake of food rich with these supplements and of products of a similar structure may have a preventive effect (inhibition of lipid peroxidation), and significantly reduce the toxic influence of these metals.

# INTRODUCTION

Malondialdehyde (MDA) is the final product in the lipid peroxidation process. The degree of lipid peroxidation, oxidative stress, can be evaluated by measuring the level of MDA in different tissues.

Heavy metals (Cd, Pb, Cu) represent a serious ecological problem due to their solubility, mobility and ability to accumulate in the soil. The exposure to these heavy metals can originate from different sources (drinking water, food, air), and they can make their way into the human body through the respiratory and digestive system.

The toxicity of these heavy metals is also manifested in the stimulation of the release of free radicals and reactive oxygen species in the body. This triggers oxidative stress and leads to lipid peroxidation of cell membranes, which disrupts their functionality and selectivity, which leads to numerous negative effects.

In this study, we determined the content of MDA, as a marker of lipid peroxidation, in the internal organs (liver, kidneys, brain and pancreas) of Wistar rats intoxicated with heavy metals. In addition, the potential protective effects of O- and S- donor supplements (glutathione and lipoic acid) on lipid peroxidation effects were also examined.

# LIPID PEROXIDATION

Lipid peroxidation is a form of oxidative damage which affects cell membranes, lipoproteins and other molecules which contain lipids, under conditions of oxidative stress [1-3]. The lipids in the cell membrane (phospholipides, glycolipids and cholesterol) are the most frequently found substrates of oxidative attacks. Lipid peroxidation is one of the best studied processes of cell damage under the conditions of oxidative stress [1]. Lipid peroxidation as a radically initiated chain reaction is a self-propagating process in the cell membrane, and is the result of the effect of free radicals on unsaturated fatty acids in the cell membranes, which leads to their damage. Free radicals are the initiators and terminators of the process of lipid peroxidation, the hydroxyl radical (OH•), as well as other radicals (for example superoxides) can trigger the oxidation process of unsaturated fatty acids.

$$HO_2• + ROOH \rightarrow RO_2• + H_2O_2 \ [4]$$

Lipid peroxidation has three phases: the initiation, propagation and termination. The initiation of the process of lipid peroxidation includes the subtraction of hydrogen atoms by radicals originating from oxygen with a high oxidation potential (OH and $O_2$), from the metyl group ($-CH_2-$) in the $\alpha$ position in relation to the C=C relationship in the carbohydrate chain of unsaturated fatty acids, Figure 1.

In order to initiate the lipid peroxidation and the Fenton reaction, the presence of $Fe^{3+}$ and $Fe^{2+}$ ions are needed at a ratio of 1:1. Lipid peroxidation decreases the fluidity of the biological membranes which results in increased permeability for univalent and bivalent ions, water and the inactivation of the membrane enzymes [5]. The damage to the lysosome membrane leads to the emergence of hydrolytic enzymes, while the disruption of the membrane structure of the mitochondria leads to the release of $Ca^{2+}$ and the activation of $Ca^{2+}$-dependent enzymes. The lipid peroxidation process decreases the hydrophobic effect of the lipid bylayer, which changes the affinity and interaction between the proteins and the lipids, the essential components of the process of separation of a significant number of membrane proteins such as: endocytosis, phagocytosis, exocytosis, and ion transport [6-8].

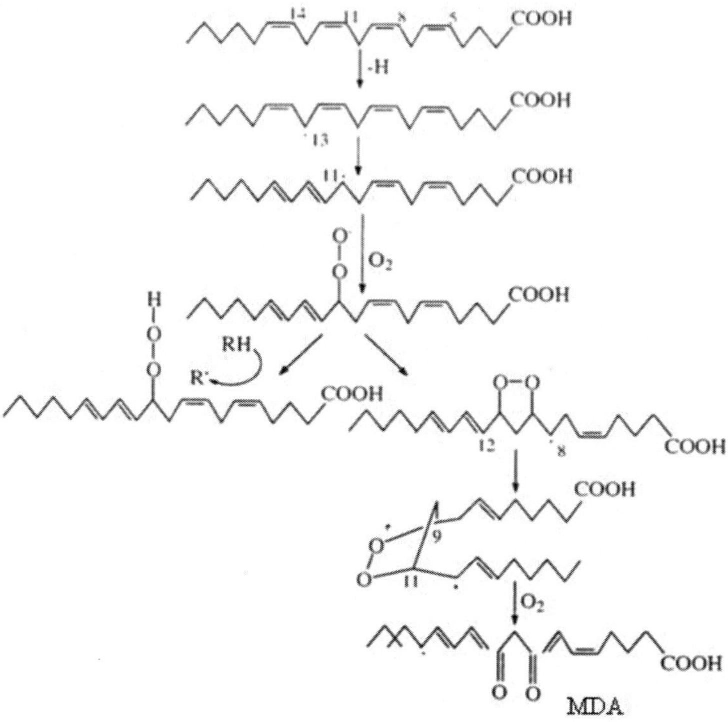

Figure 1. The initiation and propagation of the process of lipid peroxidation [1].

The fragmentation of the chain of fatty acids up until the intermediary tip of aldehydes and short-chain evaporable carbohydrates leads to the loss of membrane integrity, while the rupture in lysosome membranes leads to the release of hydrolytic enzymes which further damage the cell [9]. The process of lipid peroxidation initiates cell death [6, 10]. The toxic effects of this process are always present in the conditions of an inadequate balance between pro-oxidative and anti-oxidative cell factors [11].

During the process of lipid peroxidation, the primary highly reactive intermediaries are formed: alkyl radicals, conjugated dienes, peroxy- and alcosy-radicals, as well as lipid hydroperoxides, as is illustrated in Figure 2. Through the further decomposition of these primary products, the secondary products of lipid peroxidation emerge: evaporable carbohydrates, aldehydes, 4-hydroxy- 2,3-*trans*-nonenal and 4,5-dihydroxydecenal and the end product of lipid peroxidation, malondialdehyde.

Metals with an altering valence can initiate the process of lipid peroxidation by creating various reactive oxygen species [12].

Secondly, when the metal ions are added to the lipid systems which already contain peroxides, they transform the peroxides into peroxyl and alcoxy radicals which can further add hydrogen, whereby propagating the process of lipid peroxidation. The end product of the effect of metal on lipid peroxide is the cytotoxic MDA, 4-hydroxynonenal and the gasses ethane and pentane [13-15].

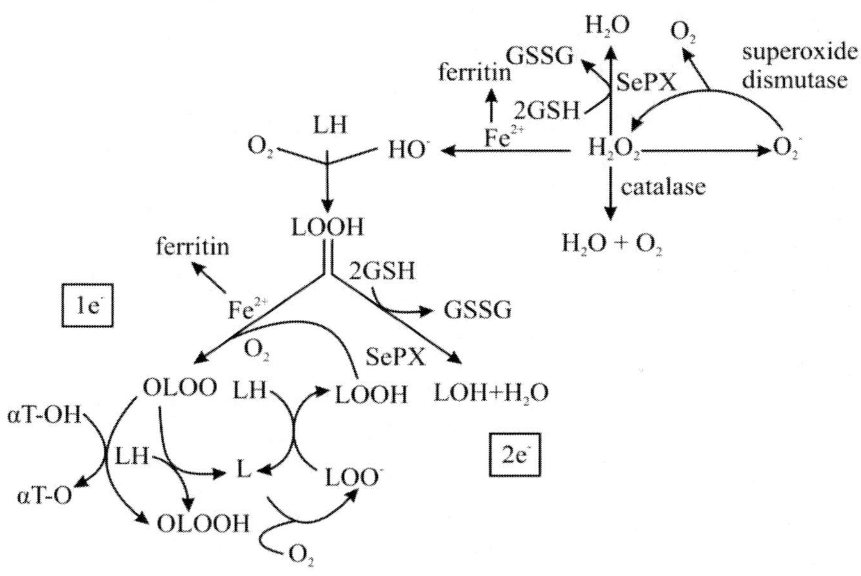


Figure 2. The emergence, metabolism and detoxication of lipid hydroperoxides [16].

Lipid peroxidation is the most pronounced negative mechanism in the effect of the reactive oxygen species (ROS), which ends in irreversible damage to the functions and structure of the cell membrane of humans, animals and plants. The end products of the indicators of the levels of lipid peroxidation are the thiobarbituric acid reactive substances (TBARS).[17]. In this process, the toxic metals which contribute to the production of reactive oxygen species and block –SH active groups of enzymes, along with the possible interactions via their –S donor atoms play an indirect role, which also leads to the formation of coordinately-covariant structures of the –S–M type.

## MALONDIALDEHYDE (MDA)

MDA occurs as the final product of the lipid peroxidation process (Figure 1.). This compound is a reactive aldehyde which causes oxidative stress in the cells, and is a bio-marker of the level of oxidative stress in the body or certain organs. The MDA indicates especially high cytotoxic effects. The reaction between the MDA and the proteins, RNA, DNA or phospholipids can lead to a modification of these substances and to damage to the cell membranes, and intracellular molecules [18]. The end product of lipid peroxidation (MDA), once it is bound to the DNA, creates the so-called "DNA radicals" which are responsible for the emergence of mutations. MDA is a reactive potential mutagen and carcinogen. The MDA also inhibits numerous thiol-dependent enzymes: glucose-6-phosphatase, $Na^+K^+$-ATPase, adenylate cyclase, $Ca^{2+}$-ATPase [3]. The extent of the lipid peroxidation is estimated by measuring the levels of MDA in various tissues [19].

In addition to the increase in the concentration of the MDA, the presence of pentane can also be registered, for example in the expired air of individuals suffering from rheumatoid arthritis [20], acute myocardial infarction [21], multiple sclerosis [22], and respiratory distress syndrome [23]. The increased elimination of ethane is noted after a full [24], and among patients with colitis [25].

MDA and other aldehydes can react with amino-groups, forming intra-molecular covalent and hydroxyl bonds, thus changing their structural-functional features [26]. The products of lipid peroxidation are easily detected by means of a spectrophotometric method and are used to measure the level of oxidative stress, usually to determine the concentration of MDA [27]. Increased concentrations of MDA indicate the presence of oxidative stress due the disruption of the balance between the prooxidative and antioxidative system.

## ANTIOXIDANTS IN BIOLOGICAL SYSTEMS

Antioxidants represent all the substances which, present in a small concentration in relation to the substances which are being oxidized, can either prevent or significantly decrease their oxidation [28]. From a functional point of view, the antioxidative protection of the human body includes three levels of action. According to the nature and manner of functioning, antioxidants can be divided into enzyme and non-enzyme. Enzymes (superoxide-dismutase, catalasys, glutathione perioxidase, glutathione reductase, glutathione S-transferase) make up the so-called first line of antioxidative protection which completely prevents the endogenic formation of free radicals. The non-enzyme antioxidants represent the secondary line of defense which includes the engagement of systems under the conditions of normal and heightened formation of free radicals [11].

The non-enzyme antioxidants include: vitamin E, vitamin C, thiol compounds (glutathione, lipoic acid, methionine, cysteine), dihydrolipoic acid, albumin, metallothionein, bilirubin, uric acid, estrogen, creatinine, coenzyme Q, polyamine, β-carotene, flavonoids and other phenolic compounds of plant origin.

Pharmacologically active substances include: nonsteroid anti-inflammatory medication, calcium blockers, allopurinol, N- acetylcysteine, angiotensin-converting-enzyme (ACE) inhibitors and deferoxamine, and also enable non-enzyme antioxidative protection via various mechanisms [29].

The third level of antioxidative protection is realized by antioxidant enzymes which take part in the reparation of the emerging oxidative lipid damage, protein damage, carbohydrate damage and nucleic acid damage. The enzymes responsible for the reparation and removal of oxidative substrates include: endo- and exo-nucleases, DNK-ligase, DNK-polimerase, classic and phospholipid-dependent glutathione peroxidase, phospholipase A2, various proteolytic enzymes, methionine sulfoxide reductase, glycolysis and others [30-33].

## Glutathione (GSH)

GSH is a tripeptide L-$\gamma$-glutamyl-cysteinyl-glycine, which makes up 90% of the overall non-protein sulfate compounds in the cell, and is an essential cofactor of some enzymes (glutathioneperoxidase, glutathione S-transferase, glutathione transhydrogenase, glutathione reductase). GSH is a biological redox agent (Figure 3) in erythrocyte metabolism and plays a role in the transfer of amino acids. It is widespread in human and animal tissues, and plants and microorganisms. Intra-cell concentrations of 0.1–10 mM make GSH one of the most frequent thiol compounds [34, 35].

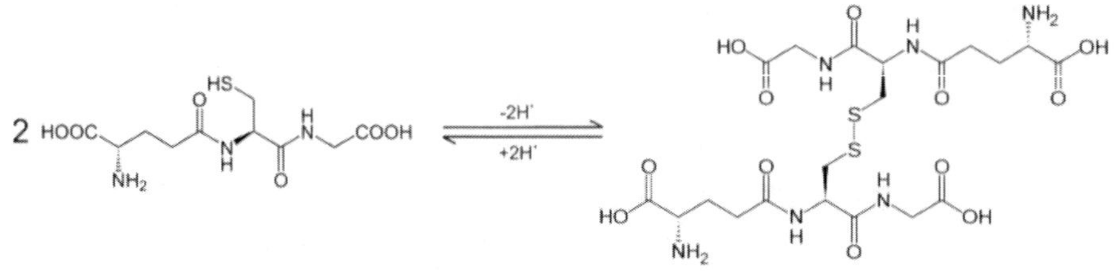

Figure 3. Redox system GSH-GSSG.

GSH as a thiol compound is an antioxidant found within the cell. Based on the results obtained upon defining the content of GSH in mosquitoes, flies, mice, rats and humans, a hypothesis was formed whose concentration reduces with age, and which is the possible key to ageing and the appearance of different pathological states.

It is widespread in human and animal tissue, plants and microorganisms. The main sources of glutathione in food include: broccoli, spinach, avocados, Brussel sprouts, cauliflower, cabbage and kale. Glutathione as a thiol compound functions as an antioxidant in the cell. This tri-peptide is found in high concentrations in almost all cells. It can be found in the cytoplasm, nucleus, and mitochondria and is the main soluble antioxidant in all the cell parts. Glutathione is the main non-protein thiol. Glutathione can only be found in aerobic bodies [36].

In the cell, glutathione can take two routes: the cytoplasmic and mithochondrial, which it reaches via cytoplasm. The intracellular concentration of glutathione is regulated via the activity of the enzymes included in the synthesis, and then via the available amino acids, especially cysteine, the intensity of the expenditure in the detoxication process in the cell and the redistribution of GSH between the organs [37, 38].

Glutathione functions as a reducing agent in numerous enzyme and non-enzyme reactions. It is a co-substrate in the enzyme reaction which is a catalyst for glutathione peroxidases, as a reaction during which GSSG emerges. Oxide glutathione, with the help of glutathione reductase, once again regenerates GSH in the presence of NADPH (Figure 4).

Due to the presence of a reactive sulfhydryl group from the molecule, glutathione belongs to the basic participants in the cell antioxidative system. The sulfur atom in the sulfhydryl group is easily adjusted to the loss of one electron, and the life span of the thiol radical can significantly be longer than the life span of other free radicals.

*oxidation:* $O_2^- \bullet + H^+ + GSH \rightarrow GS\bullet + H_2O_2$
*ionization:* $GSH \rightarrow GS^- + H^+$

Under physiological conditions for a certain pH level, sulfhydryl groups can also partially be ionized, creating thus a thiolate anion which is more reactive nucleophile (it is responsible for the reaction of the thiol in the metabolism of xenobiotics). The reactions of sulfhydryl groups during oxidative stress include cases in which the sulfur radicals and thiolate anions play an important role.

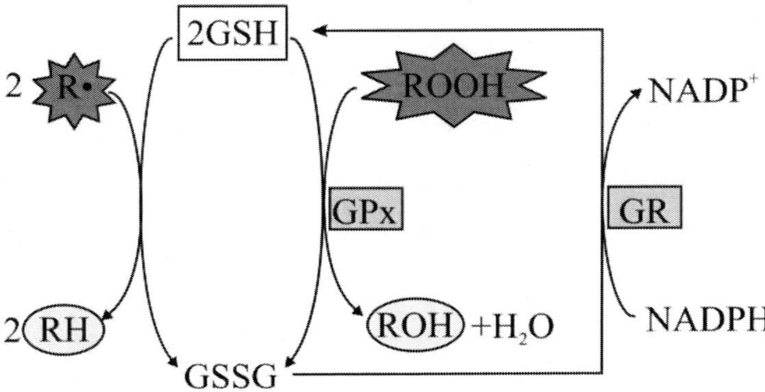


Figure 4. A schematic image of the function of GSH as an antioxidant.

The antioxidative role of glutathione can also be seen in the slowing of the aging process, mutagenesis, and carcinogenesis. Glutathione takes part in the regulation of various other processes, for example the detoxication of xenobiotics, the synthesis of eicosanoids, the synthesis of nucleic acids and proteins, in cell signalization, and proliferation and differentiation. The level of GSH decreases with time due to the inclusion of partially hydrogenated vegetable fats and the excessive exposure to toxic substances, drugs and pesticides [17, 35, 39].

## Lipoic acid (LA)

α-Lipoic acid ((R)-5-(1,2-dithiolan-3-yl)pentanoic acid) is an organic compound occurring naturally in small quantities in plants and animals as well as in humans [40]. Its content in human serum is about 16 mg/L [41]. Lipoic acid contains two sulfhydryl groups which may be in either an oxidized or reduced state (Figure 5). The reduced form is called dihydrolipoic acid (DHLA) while the oxidized form is usually referred as lipoic acid (LA) [42].

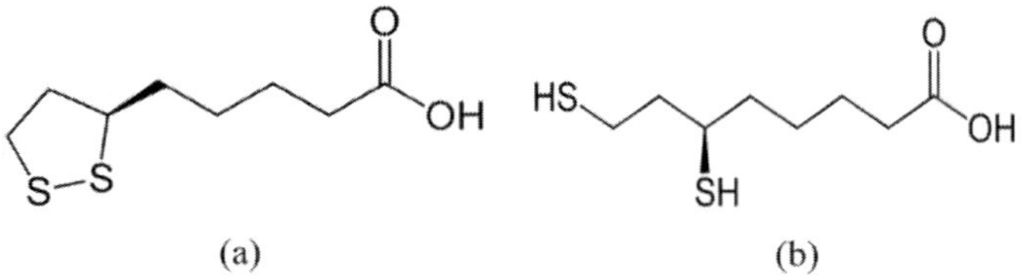

Figure 5. The oxidized (a) and reduced (b) form of α-lipoic acid (LA).

LA can be found in various foods covalently bound to the lysine in proteins (lipoyllysine). Although it was found in foodstuffs of both plant and animal origin, quantitative data about the content of LA or lipoyllysine are limited. Animal organs rich in lipoyllysine ($\sim$1–3 µg/g) are the kidneys, heart, and liver. Plant species containing lipoyllysine are usually the ones rich in chlorophyll, like spinach and broccoli [43]. LA ingested by food is activated with ATP or GTP by a lipoate-activating enzyme and then transferred to LA-dependent enzymes by lipoyl transferase [44]. Oral administration of high doses of LA (50 mg and more) causes a significant, but temporary, increase of free LA in plasma and cells. Pharmacokinetic studies in humans showed that approximately 30–40% of orally administrated LA was absorbed [45]. LA is rapidly reduced to DHLA in cells, and in vitro studies showed its quick elimination [46]. Endogenous synthesized LA is covalently bound to specific proteins acting as cofactors in some important enzymic complexes. Beside its physiological role in protein-bound form, there is growing interest in the potential therapeutic use of pure LA [47] with a special emphasis on its antioxidant activities [48-51] and beneficial effects in heavy metal intoxication [52-57].

# METAL TOXICITY

Cadmium, lead and copper represent a serious ecological problem due to their soluble nature, their mobility and ability to accumulate in the soil. Exposure to these heavy metals can originate from different sources (drinking water, food and air), and they can enter the human body through the respiratory and digestive systems. Through accumulation in the internal organs, they manifest a hepatotoxic, nephrotoxic and neuroacoustic effect [58, 59].

The toxicity of heavy metals is manifested through their reaction with – SH groups of the protein part of the molecule, which leads to a change in the activity of the enzyme, but also the glutathione.

$$\text{Protein-SH} + M^{n+} \rightarrow (\text{Protein–S})_m M + mH^+$$

## Lead

Lead belongs to the group of the highly toxic elements, with a cumulative-toxic effect [60]. Lead is not an essential metal, but is present in all of the tissues and organs of mammals. It can mostly be found in mineral tissue – bones and teeth (over 90% of the overall amount of this element) [61]. If constantly introduced into the human body, even in small amounts, lead partially replaces calcium in the tertiary calcium-phosphate bone skeleton, where its toxic effect is gradually increased. Daily amounts of lead which a human normally absorbs mostly through food and drink, can range up to 0.3 mg. However, this amount does not cause poisoning, since lead is excreted in approximately the same daily amount from the human body [62]. Lead intake can occur in different ways. Lead bound in tetraethyl lead, as an addition of gasoline, through its combustion is transferred into the atmosphere and reaches the human respiratory system. Part of the lead is absorbed by plants and animals alike, so it is introduced into the human body by means of food. The innards used in the human diet (especially the liver and kidneys) contain high concentrations of lead. Nevertheless, it has been proven that only 3% the lead in the innards is absorbed into the human body. Besides food, lead can be introduced into the body by means of water, which lead reaches via the air, soil or pipelines. Studies have shown that lead from water or other beverages are reabsorbed to a greater extent than the lead from food. In addition, lead introduced into the body between meals is absorbed to a greater extent than the lead introduced during a meal, while a greater frequency of food intake minimizes the absorption of lead. It has been proven that 50% of lead is absorbed from water, following an overnight fast [60]. Exposure to lead in the workplace, in factories and workshops leads to severe and prolonged illnesses as the gravest of professional illnesses. According to the American Center for Disease Control (CDC) [62], a lead content in the blood of less than 240 $\mu g/dm^3$ is *normal*, 250-490 $\mu g/dm^3$ belongs to the *moderate risk* category, 500-

690 µg/dm$^3$ to the *high risk* category and amounts that exceed 700 µg/dm$^3$ fall into the *urgent risk* category. Nevertheless, prolonged exposure to low-level toxicity (< 240 µg/dm$^3$) can lead to various psychological disorders and learning disabilities among children. Naturally, these symptoms can occur among children even in the case of intake of lead amounts of less than 50 µg/dm$^3$ [63].

Approximately 99% of lead in the blood is bound to erythrocytes since these blood cells have a large affinity to bind to heavy metals, especially to lead, resulting in erythrocytes being very sensitive to oxidative damage and lipid peroxidation [64].

Lead has a high affinity for the S–binding locations in the tissues, and easily builds indissoluble sulfides and reacts with parts of the biomolecules and free –SH groups [34].

Lead contributes to the formation of reactive oxygen species *in vivo*, which leads to the decrease in internal antioxidative defenses, and causes disorders in the exchange of ion electrolytes through the cell membrane [65]. At the molecular level, the effect of lead can be seen in the increased production of reactive oxygen species and the stimulation of lipid peroxidation [66, 67].

## Cadmium

Cadmium is one of the most dangerous poisons of the working and living environment, which can be introduced into the body by means of air, food or drinking water. Thus, the lack of iron can significantly increase the accumulation of cadmium, and sufficient amounts of iron in the blood inhibit the accumulation of cadmium. In addition, increased doses of vitamin D act as an antidote to cadmium poisoning [34]. The cadmium content in the human body has a value of $1\times10^{-4}$ % of the overall body mass [68]. Cadmium poisoning can be acute and chronic. Acute poisoning occurs due to inhalation of the fumes of particles of cadmium salts (oxides, chlorides, sulfides, sulfates, carbonates and acetates) whose concentration in the air is approximately 1mg/m$^3$ [69]. The toxic effect of cadmium to a great extent depends on the intake of calcium. Low calcium intake leads to higher cadmium absorption, the retention, accumulation and increased toxic effect of this metal. The consequences of this include kidney and bone damage (osteomalacia), as well as hypertension and anemia [70].

Once it enters the body, cadmium is transported into the blood by means of red blood cells and a highly molecular blood protein – albumin. The normal level of cadmium in the blood of adults is less than 1 µg/dm3. Even though cadmium circulates via the blood throughout the entire body, the greatest accumulation (from 50 to 60% of the body's cadmium load) can be found in the kidneys and liver [71].

Cadmium bound to metallothionein is accumulated in the liver, kidneys, pancreas, and in smaller amounts in other tissues as well. After 24 hours, the concentration of cadmium in most tissue types remains constant, except in the kidneys where it gradually increases.

As is the case with other metals, cadmium also participates very slightly or not at all in direct metabolic exchange, but is bound to various biological components such as proteins, thiol (-SH) groups and anion groups of various macromolecules [34].

Even though cadmium is not a redox-active metal and cannot directly participate in the Fenton reaction and the creation of reactive oxygen species (ROS), through indirect mechanisms it leads to oxidative stress [72]. Cadmium causes oxidative damage to the DNA, proteins and lipids, which is probably the influence of Cd on the enzyme and non-enzyme components of the antioxidative system of defense of the human body. In addition, there is evidence that cadmium increases the level of lipid peroxidation [73, 74].


## Copper

Copper is a biogenic element found in numerous enzymic systems and also plays an important role in suppressing inflammatory processes. The human body contains approximately 80–120 mg of copper under normal physiological conditions, and its normal concentration range in the blood plasma is 70–150 µg/L. Changes in the level of copper outside the normal ranges is connected with many diseases like aceruloplasminemia, Wilson's disease, Menkes disease, Alzheimer's disease, and others. The increased intake of copper may cause its accumulation in some tissues and organs and chronic intoxication may cause a disorder in some metabolic pathways, a reduction of free radical species and even apoptosis. It is well-known that Cu(II) ion ($d^9$ configuration) can give coordination compounds of various coordination and geometry through O-, N-, and S-donor atoms of different biomolecules and small molecules. This fact, together with its redox activity, is responsible for various roles copper can play in living organisms and possible interactions with many antioxidant nutrients, including lipoic acid, which are of special importance [75-77].

The therapy used to remove toxic heavy metals from the human body is the so-called chelation therapy It is based on the coordination ability of metals and the donor abilities of the O, N atoms and chelating agents, ligands which are used to form stable associate types of complex particles and thus eliminate the negative effect of free metal ions. The emerging products can via bodily fluids be eliminated from the body. "Good" chelators are, among other substances and vitamins, glutathione and lipoic acid [77-81].

The effects of heavy metal intoxication (Cd, Pb, Cu) and the protective role of the supplements have been studied through the measurement of the concentration of MDA on the model system of experimental animals, albino Wistar rats, all females, aged two months, which were bred in laboratory conditions of the Vivarium of the Faculty of Medicine, University of Niš (Serbia). They were divided into twelve groups [53].

# MDA UNDER THE CONDITIONS OF CHRONIC LEAD INTOXICATION

The consequences of chronic lead(II) poisoning (*total dose of 12 mg/kg of Pb in the form of lead(II)-acetate in saline for a period of 3 weeks by intraperitoneal injection*) monitored via the MDA as indicators of the level of lipid peroxidation were studied on a model system of Albino Wistar rats. The measurements of the levels of MDA in homogenates of the liver tissue, kidney tissue, brain and pancreas are shown in Figure 6, compared to the control group which adhered to a normal diet and lifestyle.

The level of MDA in the liver significantly increased (approximately six times) from 0.45±0.14 μmol/mg for the control group to 2.48±0.23 μmol/mg for the group exposed to the influence of lead, and approximately three times in the other organs: in the kidneys from 0.48±0.06 μmol/mg to 1.49±0.38 μmol/mg; in the brain 1.01±0.10 μmol/mg to 2.72±0.35 μmol/mg and in the pancreas 0.75±0.18 μmol/mg to 2.34±0.32 μmol/mg.

The toxicity of the lead interacts with the cell membranes via unsaturated fatty acids, by disrupting the peroxidative-antioxidative balance, which indirectly leads to the increase in the level of the final product of lipid peroxidation [66, 82]. By applying the glutathione one day following poisoning, the level of TBARS diminishes by almost 50% in the liver, kidneys and pancreas. The level of MDA in the brain remains significantly high, which correlates with the results of, and indicates that lead poisoning leads to oxidative damage to the nervous system.[83].

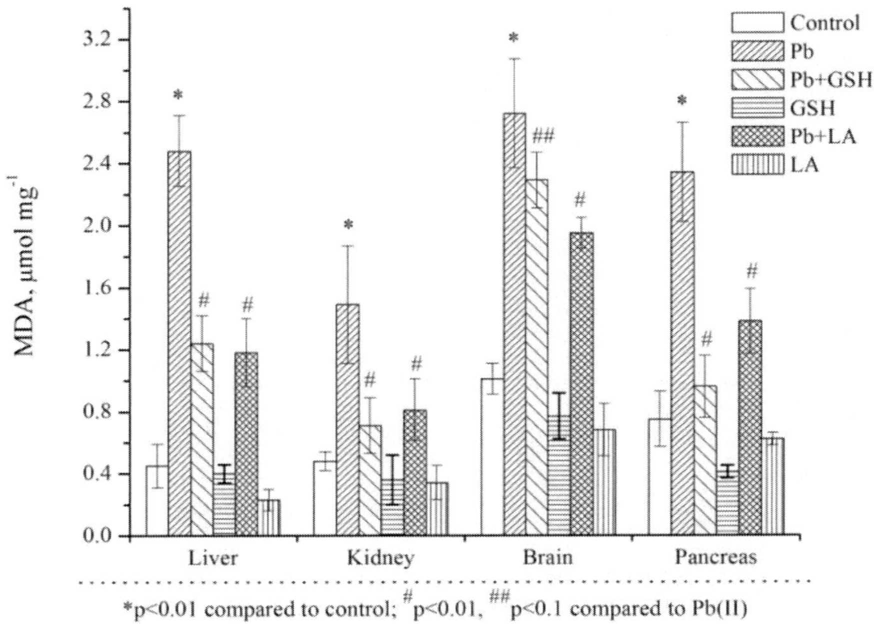


Figure 6. MDA levels in liver, kidney, brain and pancreas tissue homogenates of experimental animals under the conditions of Pb(II) intoxication.

Lead has a pronounced affinity for binding with sulfur and building mercaptides. This ion blocks the –SH reaction places of the protein, enzyme, and antioxidants. With the addition of the antidote, thiol compounds such as lipoic acid and glutathione with which the $Pb^{2+}$ ion can react, decrease its toxicity. It is likely that products rich in natural thiol compounds can also be good supplements, useful under the conditions of chronic exposure to lead.

The results of this study have indicated that supplementation via lipoic acid significantly reduces the level of lipid peroxides, that is, the value of MDA. This indicates the reduction in the level of oxidative damage to the tissue of the liver, kidneys, pancreas and brain [84, 85].

## MDA UNDER THE CONDITIONS OF ACUTE CADMIUM INTOXICATION

Figure 7. shows the levels of MDA in the homogenates of the organs of Wistar rats after acute cadmium poisoning (*total dose of 0.56 mg/kg of Cd in the form of cadmium(II)-chloride in saline over a period of 5 days by subcutaneous injection*) [53].

The level of MDA in comparison to the control group of animals increased in the liver from 0.45±0.14 μmol/mg to 2.20±0.22 μmol/mg among animals intoxicated with this metal; in the kidneys 0.48±0.06 μmol/mg to 1.83±0.19 μmol/mg; in the brain 1.01±0.10 μmol/mg to 3.08±0.46 μmol/mg and in the pancreas 0.75±0.18 μmol/mg to 2.92±0.43 μmol/mg. The indicator of the levels of lipid peroxidation TBARS significantly increased, three to four times. This is probably a consequence of the exhaustion of physiological antioxidants, especially those with –SH groups, with which cadmium indirectly reacts [74].

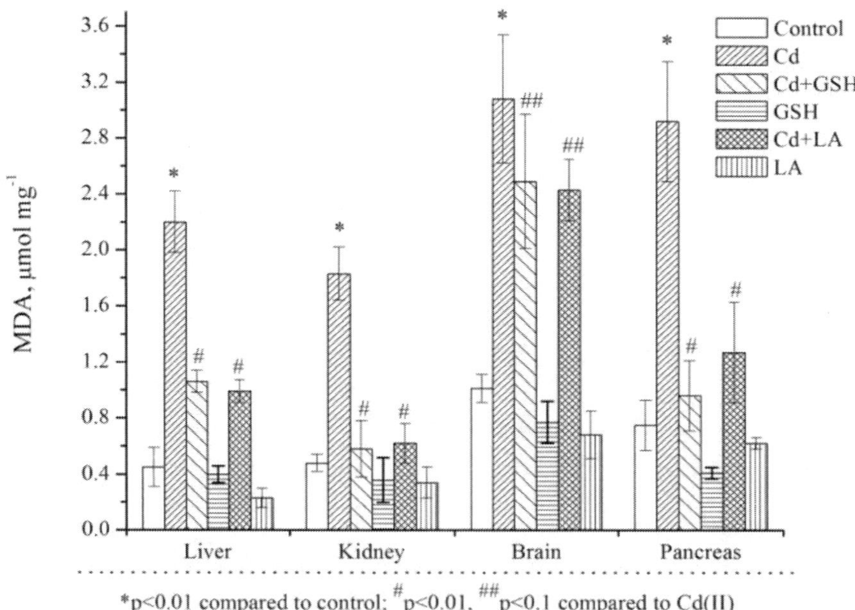


Figure 7. MDA levels in liver, kidney, brain and pancreas tissue homogenates of experimental animals under the conditions of Cd(II) intoxication.

By applying glutathione the day after the cadmium poisoning (Cd:GSH = 1:2) we can decrease the level of TBARS in the liver and brain (2.20 ± 0.22 μmol/mg to 1.06 ± 0.08 μmol/mg in the liver, from 3.08 ± 0.46 μmol/mg to 2.49±0.48 μmol/mg in the brain) and almost completely eliminate the effects of cadmium in the kidneys and pancreas (from 1.83±0.19 μmol/mg to 0.58 ± 0.20 μmol/mg in the kidneys, from 2.92 ± 0.43 μmol/mg to 0.96 ± 0.25 μmol/mg in the pancreas).

By applying lipoic acid one day after poisoning in the mole ratio of Cd:LA=1:2, the level of MDA in the liver is normalized (from 2.20 ± 0.22 to 0.99 ± 0.08 μmol/mg), while in the other organs it is reduced to approximately half its size (from 1.83±0.19 to 0.62±0.14 μmol/mg in the kidneys, from 3.08 ± 0.46 to 2.43 ± 0.22 μmol/mg in the brain, and from 2.92 ± 0.43 to 1.27 ± 0.36 μmol/mg in the pancreas).

## MDA under the Conditions of Chronic Copper Intoxication

The results of MDA measurements in tissue homogenates of rats intoxicated with Cu(II) ion (*total dose of 10.8 mg/kg of Cu in the form of copper(II)-sulfate in saline for a period of 3 weeks by subcutaneous injection*) with and without LA supplementation are shown in Figure 8. As can be seen from Figure 8., Cu(II) ion intoxication caused a statistically significant ($p < 0.01$) increase of MDA levels in both the liver and kidney tissues from 0.45 ± 0.14 to 1.67 ± 0.35 μmol/mg and from 0.48 ± 0.06 to 1.00 ± 0.39 μmol/mg, respectively. LA supplementation one day after intoxication of animals with Cu(II) ion reduced its toxic effects significantly. MDA concentration decreased by approximately 40% (to 1.02 ± 0.26 μmol/mg) in liver homogenates and by approximately 25% in kidney homogenates (to 0.74 ± 0.28 μmol/mg). LA supplementation without Cu(II) ion intoxication also reduced MDA levels in both the liver (about 50% to 0.23 ± 0.07 μmol/mg) and kidney tissues (approximately 30% to 0.34 ± 0.11 μmol/mg) in comparison to animals in control group. These findings are consistent with the fact that copper, due to its redox activity, may induce oxidative damage in various tissues by catalyzing the production of free radicals and other reactive oxygen species (ROS) and reactive nitrogen species (RNS) [75].

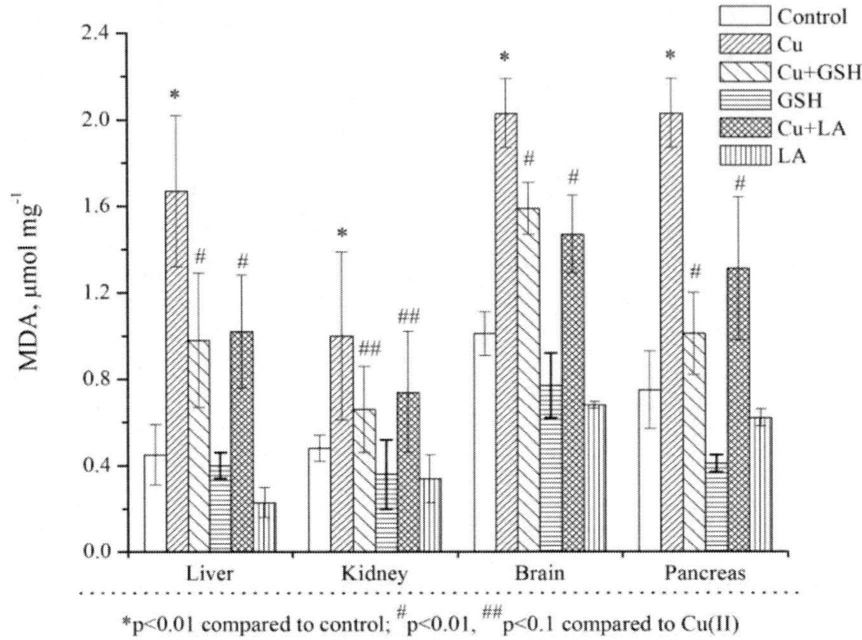


Figure 8. MDA levels in liver, kidney, brain and pancreas tissue homogenates of experimental animals under the conditions of Cu(II) intoxication.

In the cases of severe intoxication, even cell apoptosis and tissue necrosis may occur [86]. Our results clearly indicate that supplementation by LA one day after intoxication greatly reduced the toxic effects of the Cu(II) ion as evidenced by the MDA concentration decrease. LA supplementation also decreased the MDA concentrations in tissues of animals not intoxicated with the Cu(II) ion in comparison to the control group which is an indicator of its general antioxidant activity. The difference of total copper concentrations in tissue homogenates of rats intoxicated with Cu(II) ions and intoxicated rats with subsequent LA supplementation was not statistically significant [53, 77]. However, the difference of MDA levels in tissue homogenates for these two groups of animals was significant, and this indicated the possible interaction of Cu(II) ions with LA, which may account for the protective effect of the LA.

The toxic effects of investigated metals being more pronounced in liver, respectively by metal: Pb > Cd > Cu. In other organs the toxic effect of metals, according of the level of MDA is respectively: Cd > Pb > Cu.

# PROTECTIVE ROLE OF GSH AND LA

From obtained results (Figure 6, 7 and 8) we can see the benefit effects that glutathione and lipoic acid have in lead poisoning, but also cadmium and copper poisoning. The application of these supplements during intoxication decreases the negative effects measured via the level of the MDA, the indicator of lipoic peroxidation.

Glutathione as a chelate agent contains active sites, including the –SH group which is used to form mercaptides with heavy metal ions (Ag, Cd, Pb, Cu). It can form complex association types $[M(GSH)_2]$. Their stability depends, among other things, on the size of the ion, acid-base features, electronic configurations, and so the affinity of the metals towards the thiol group decreases in the following order $Cd > Pb > Cu$.

Glutathione takes part in the reduction of disulfide and other molecules, as well as in conjugation with them. That is how it protects the cells from oxidative damage due to free radicals. Glutathione decreases free radicals through the non-enzyme reaction of $H_2O_2$ reduction, and other hydroperoxides whose production is aided by heavy metals.

The level of GSH in various organs is determined by the dynamics of inter-organic redistribution of GSH and its intermediaries [38]. The ability to excrete GSH from various organs, especially the liver determines the higher concentration of GSH in the interstitium in comparison to the concentrations in the plasma, which represents an essential factor in the protection of the cell membrane from the oxidants in its immediate environment [37, 87]. The excretion of the GSH into the peripheral bloodstream is enables through the use of precursors for their synthesis. Considering that the liver is a significant source of GSH for other peripheral tissues, the intense metabolism of xenobiotics in the human body, which leads to a decrease in the concentration of results in the decrease in the concentration of GSH in other peripheral tissues.

The kidney represents the primary organ which enables the extraction of GSH from the peripheral bloodstream. It is in the kidneys that the GSH is actively synthesized and secreted. Under the conditions of intense oxidative stress, the excretion of GSH from the liver into the peripheral bloodstream is more intense This is how the availability of the GSH is provided for other organs [88, 89].

In the human body, the lipoic acid is found both intracellularly, extracellularly and on cell membranes. Lipoic acid as a hydrosoluble compound is a good antioxidant which easily permeates into the cytoplasm and manifests its effects there. Obtained results show that the application, under the conditions of both chronic and acute heavy metal intoxication, significantly lowers the level of MDA in various tissues and organs. It can also establish interactions with metals via the oxygen in the carboxile group, but also through sulfur, under the condition that it is found in reduced form. The protective role of LA results in the decrease of the toxic effects of the metals in the following order: Cd > Pb > Cu.

Lipoic acid indirectly leads to an increase in the level of GSH in the cell, and this can additionally increase its effectiveness as a protector, in comparison to the glutathione.

Studies on the interaction between LA and $Cu^{2+}$ ions, at the micromolar level, have indicated that a stable complex particle is formed between them. FTIR studies have shown that the achieved coordination is $Cu^{2+}$–O–LA lipoic acid as the monodentate ligand. Lipoic acid in its reduced form emerges during interaction with heavy metals as the bidentate S-donor ligand [77].

# CONCLUSION

In our study, we have shown that the process of lipoic peroxidation is caused by heavy metal poisoning (Cd, Cu, Pb) can successfully be monitored through the concentrations of MDA in the tissue of various organs. Studies carried out on albino Wistar rats have shown that intoxication leads to a high increase in the level of MDA (with values which are two to six times greater). The toxic effects of investigated metals being more pronounced in liver, respectively by metal: Pb > Cd > Cu. In other organs the toxic effect of metals, according of the level of MDA is respectively: Cd > Pb > Cu.

By adding supplements, natural thiol compounds (LA and GSH), a significant decrease in the negative effect of intoxication is achieved, that is, the process of lipid peroxidation. The protective role of GSH is manifested in the reduction of negative effects of intoxication in the following order of the metals Cd > Pb > Cu. The protective role of LA results in the decrease of the toxic effects of the metals in the following order: Cd > Pb > Cu. Both protectors manifest the best effect by eliminating the effect of the $Cu^{2+}$ ion in the brain.

## ACKNOWLEDGMENTS

This work was supported by the Ministry of Education and Science of the Republic of Serbia [grant number TR31060]. The authors would also like to thank Marta Dimitrijević, MA, for proofreading the text of this paper.

## REFERENCES

[1]   Halliwell B, Gutteridge J. Oxygen toxicity, oxygen radicals, transition metals and disease. *Biochemical journal.* 1984;219(1):1.

[2]   Kagan VE. Lipid peroxidation in biomembranes. *Boca Raton: CRC;* 1988.

[3]   Slater TF. Free radical mechanisms in tissue injury. Cell Function and Disease: Springer; 1988. p. 209-18.

[4]   McCord JM. Superoxide radical: controversies, contradictions, and paradoxes. *Experimental Biology and Medicine.* 1995;209(2):112-7.

[5]   Bruch RC, Thayer WS. Differential effect of lipid peroxidation on membrane fluidity as determined by electron spin resonance probes. Biochimica et Biophysica Acta (BBA) - *Biomembranes.* 1983;733(2):216-22.

[6]   Nicolson GL. Transmembrane control of the receptors on normal and tumor cells: I. Cytoplasmic influence over cell surface components. *Biochimica et Biophysica Acta (BBA) - Reviews on Biomembranes.* 1976;457(1):57-108.

[7]   Bellomo G, Orrenius S. Altered thiol and calcium homeostasis in oxidative hepatocellular injury. *Hepatology.* 1985;5(5):876-82.

[8]   Thomas CE, Reed DJ. Current status of calcium in hepatocellular injury. *Hepatology.* 1989;10(3):375-84.

[9]   Fong K-L, McCay PB, Poyer JL, Keele BB, Misra H. Evidence that peroxidation of lysosomal membranes is initiated by hydroxyl free radicals produced during flavin enzyme activity. *Journal of Biological Chemistry.* 1973;248(22):7792-7.

[10]  Korsmeyer SJ, Yin X-M, Oltvai ZN, Veis-Novack DJ, Linette GP. Reactive oxygen species and the regulation of cell death by the Bcl-2 gene family. *Biochimica et Biophysica Acta (BBA)-Molecular Basis of Disease.* 1995;1271(1):63-6.

[11] Halliwell B, Gutteridge J, Cross C. Free radicals, antioxidants, and human disease: where are we now? *The Journal of laboratory and clinical medicine.* 1992;119(6):598.

[12] Aruoma OI, Halliwell B, Butler J, Hoey BM. Apparent inactivation of α1-antiroteinase by sulphur-containing radicals derived from penicillamine. Biochemical Pharmacology. 1989;38(24):4353-7.

[13] Steinberg D, Witztum JL. Lipoproteins and atherogenesis: current concepts. *Jama.* 1990;264(23):3047-52.

[14] Traystman RJ, Kirsch JR, Koehler RC. Oxygen radical mechanisms of brain injury following ischemia and reperfusion. *Journal of Applied Physiology.* 1991;71(4):1185-95.

[15] Weitzman SA, Gordon L. Inflammation and cancer: role of phagocyte-generated oxidants in carcinogenesis. *Blood.* 1990;76(4):655-63.

[16] Girotti AW. Lipid hydroperoxide generation, turnover, and effector action in biological systems. *Journal of lipid research.* 1998;39(8):1529-42.

[17] Pavlović D. Free radicals, lipid peroxidation and antioxidant protection. Biochemistry. *Beolgrade: Savremena administracija* 2006. p. 692-702.

[18] Sreejayan N, von Ritter C. Lipid peroxidation in bile: the role of hydrophobic bile acids and the effect on biliary epithelial cell function. *Pathophysiology.* 1999;5(4):225-32.

[19] Yiin S, Lin T. Lead-catalyzed peroxidation of essential unsaturated fatty acid. *Biological trace element research.* 1995;50(2):167-72.

[20] Humad S, Zarling E, Clapper M, Skosey JL. Breath pentane excretion as a marker of disease activity in rheumatoid arthritis. *Free Radical Research.* 1988;5(2):101-6.

[21] Weitz Z, Birnbaum A, Skosey J, Sobotka P, Zarling E. High breath pentane concentrations during acute myocardial infarction. *The Lancet.* 1991;337(8747):933-5.

[22] Toshniwal PK, Zarling EJ. Evidence for increased lipid peroxidation in multiple sclerosis. *Neurochemical research.* 1992;17(2):205-7.

[23] Drury JA, Nycyk JA, Cooke RW. Pentane measurement in ventilated infants using a commercially available system. *Free Radical Biology and Medicine.* 1997;22(5):895-900.

[24] Arterbery VE, Pryor WA, Jiang L, Sehnert SS, Foster WM, Abrams RA, et al. Breath ethane generation during clinical total body irradiation as a marker of oxygen-free-radical-mediated lipid peroxidation: a case study. *Free Radical Biology and Medicine.* 1994;17(6):569-76.

[25] Sedghi S, Keshavarzian A, Klamut M, Eiznhamer D, Zarling EJ. Elevated breath ethane levels in active ulcerative colitis: evidence for excessive lipid peroxidation. *The American journal of gastroenterology.* 1994;89(12):2217.

[26] Nair V, Cooper CS, Vietti DE, Turner GA. The chemistry of lipid peroxidation metabolites: crosslinking reactions of malondialdehyde. *Lipids.* 1986;21(1):6-10.

[27] Andreeva L, Kozhemiakin L, Kishkun A. Modification of the method of determining lipid peroxidation in a test using thiobarbituric acid. *Laboratornoe delo.* 1988(11):41.

[28] Halliwell B. How to characterize a biological antioxidant. *Free Radical Research.* 1990;9(1):1-32.

[29] Đorđević V. Biochemical oxidation. Biochemistry. Belgrade: Savremena administracija; 2000. p. 678–705.

[30] Ursini F, Bindoli A. The role of selenium peroxidases in the protection against oxidative damage of membranes. *Chemistry and Physics of Lipids.* 1987;44(2):255-76.

[31] Sevanian A, Kim E. Phospholipase A2 dependent release of fatty acids from peroxidized membranes. *J Free Radic Biol Med.* 1985;1(4):263-71.

[32] Cotgreave IA, Moldeus P, Orrenius S. Host biochemical defense mechanisms against prooxidants. *Annual Review of Pharmacology and Toxicology.* 1988;28(1):189-212.

[33] Linn S. DNA damage by iron and hydrogen peroxide in vitro and in vivo. *Drug metabolism reviews.* 1998;30(2):313-26.

[34] Jovanović JM, Nikolić RS, Kocić GM, Krstić NS, Krsmanović MM. Glutathione protects liver and kidney tissue from cadmium-and lead-provoked lipid peroxidation. *Journal of the Serbian Chemical Society.* 2013;78(2):197-207.

[35] Meister A. Glutathione metabolism and its selective modification. *J Biol Chem.* 1988;263(33):17205-8.

[36] Parke D, Piotrowski J. Glutathione: Its role in detoxification of reactive oxygen species and environmental chemicals. *Acta Polonica Tocixology.* 1996;4:1-14.

[37] Anderson ME, Meister A. Dynamic state of glutathione in blood plasma. *Journal of Biological Chemistry.* 1980;255(20):9530-3.

[38] Bannai S. Transport of cystine and cysteine in mammalian cells. *Biochimica et Biophysica Acta (BBA)-Reviews on Biomembranes.* 1984;779(3):289-306.

[39] Shan X, Aw TY, Jones DP. Glutathione-dependent projection against oxidative injury. *Pharmacology & therapeutics.* 1990;47(1):61-71.

[40] Reed LJ, DeBusk BG, Gunsalus IC, Hornberger CS. Crystalline α-lipoic acid: a catalytic agent associated with pyruvate dehydrogenase. *Science.* 1951;114(2952):93-4.

[41] Baumgartner MR, Schmalle H, Dubler E. The interaction of transition metals with the coenzyme α-lipoic acid: synthesis, structure and characterization of copper and zinc complexes. *Inorganica chimica acta.* 1996;252(1):319-31.

[42] Kramer K, Hoppe P-P, Packer L. Nutraceuticals in health and disease prevention: *CRC Press;* 2001.

[43] Durrani AI, Schwartz H, Nagl M, Sontag G. Determination of free α-lipoic acid in foodstuffs by HPLC coupled with CEAD and ESI-MS. *Food chemistry.* 2010;120(4):1143-8.

[44] Fujiwara K, Takeuchi S, Okamura-Ikeda K, Motokawa Y. Purification, characterization, and cDNA cloning of lipoate-activating enzyme from bovine liver. *Journal of Biological Chemistry.* 2001;276(31):28819-23.

[45] Teichert J, Hermann R, Ruus P, Preiss R. Plasma kinetics, metabolism, and urinary excretion of alpha-lipoic acid following oral administration in healthy volunteers. *The Journal of Clinical Pharmacology.* 2003;43(11):1257-67.

[46] Smith A, Shenvi S, Widlansky M, Suh J, Hagen T. Lipoic acid as a potential therapy for chronic diseases associated with oxidative stress. *Current medicinal chemistry.* 2004;11(9):1135-46.

[47] Bilska A, Wlodek L. Lipoic acid-the drug of the future. *Pharmacol Rep.* 2005;57(5):570-7.

[48] Biewenga GP, Haenen GR, Bast A. The pharmacology of the antioxidant lipoic acid. General Pharmacology: *The Vascular System.* 1997;29(3):315-31.

[49] Jones W, Li X, Qu Z-c, Perriott L, Whitesell RR, May JM. Uptake, recycling, and antioxidant actions of α-lipoic acid in endothelial cells. *Free Radical Biology and Medicine.* 2002;33(1):83-93.

[50] Moini H, Packer L, Saris N-EL. Antioxidant and prooxidant activities of α-lipoic acid and dihydrolipoic acid. *Toxicology and applied pharmacology.* 2002;182(1):84-90.

[51] Packer L, Witt EH, Tritschler HJ. Alpha-lipoic acid as a biological antioxidant. *Free Radical Biology and Medicine.* 1995;19(2):227-50.

[52] Lodge JK, Traber MG, Packer L. Thiol chelation of Cu 2 by dihydrolipoic acid prevents human low density lipoprotein peroxidation. *Free Radical Biology and Medicine.* 1998;25(3):287-97.

[53] Nikolic R, Krstic N, Jovanovic J, Kocic G, Cvetkovic TP, Radosavljevic-Stevanovic N. Monitoring the toxic effects of Pb, Cd and Cu on hematological parameters of Wistar rats and potential protective role of lipoic acid and glutathione. *Toxicology and industrial health.* 2013:0748233712469652.

[54] Nikolić RS, Kocić GM, Kostić DA, Nikolić NG, Jovanović JM, Krstić NS. A study of the protective role of lipoic acid in the case of acute heavy metal intoxication (Cd, Pb, Cu) through the activity of the DNase in the liver and kidneys. *Oxidation Communication.* 2014;37(4):1103–10.

[55] Ou P, Tritschler HJ, Wolff SP. Thioctic (lipoic) acid: a therapeutic metal-chelating antioxidant? *Biochemical pharmacology.* 1995; 50(1):123-6.

[56] Suh JH, Zhu B-Z, deSzoeke E, Frei B, Hagen TM. Dihydrolipoic acid lowers the redox activity of transition metal ions but does not remove them from the active site of enzymes. *Redox report.* 2004;9(1):57-61.

[57] Rooney JP. The role of thiols, dithiols, nutritional factors and interacting ligands in the toxicology of mercury. *Toxicology.* 2007;234(3):145-56.

[58] Manca D, Ricard AC, Trottier B, Chevalier G. Studies on lipid peroxidation in rat tissues following administration of low and moderate doses of cadmium chloride. *Toxicology.* 1991;67(3):303-23.

[59] Nikolić RS, Jovanović JM, Kocić GM, Cvetković TP, Stojanović SR, Anđelković TD, et al. Praćenje efekata izloženosti olovu i kadmijumu u radnoj i životnoj sredini preko parametara standardne biohemijske analize krvi i aktivnosti endonukleaza jetre. *Hem Ind.* 2011;65:403-9.

[60] Vig EK, Hu H. Lead toxicity in older adults. *Journal of the American Geriatrics Society.* 2000;48(11):1501-6.

[61] Gulson BL, Gillings BR. Lead exchange in teeth and bone--a pilot study using stable lead isotopes. *Environmental health perspectives.* 1997;105(8):820.

[62] Goyer RA. Toxic and essential metal interactions. *Annual review of nutrition.* 1997;17(1):37-50.

[63] Banks EC, Ferretti LE, Shucard D. Effects of low level lead exposure on cognitive function in children: A review of behavioral, neuropsychological and biological evidence. *Neurotoxicology.* 1996;18(1):237-81.

[64] Sivaprasad R, Nagaraj M, Varalakshmi P. Combined efficacies of lipoic acid and meso-2, 3-dimercaptosuccinic acid on lead-induced erythrocyte membrane lipid peroxidation and antioxidant status in rats. *Human & experimental toxicology*. 2003;22(4):183-92.

[65] Flora S, Flora G, Saxena G. Environmental occurrence, health effects and management of lead poisoning. Lead chemistry, analytical aspects, environmental impacts and health effects Netherlands: *Elsevier Publication*. 2006:158-228.

[66] Gurer H, Ercal N. Can antioxidants be beneficial in the treatment of lead poisoning? *Free Radical Biology and Medicine*. 2000;29(10):927-45.

[67] Patrick L. Lead toxicity part II: the role of free radical damage and the use of antioxidants in the pathology and treatment of lead toxicity. *Alternative medicine review: a journal of clinical therapeutic*. 2006;11(2):114-27.

[68] Danielsson L-G, Jagner D, Josefson M, Westerlund S. Computerized potentiometric stripping analysis for the determination of cadmium, lead, copper and zinc in biological materials. *Analytica Chimica Acta*. 1981;127:147-56.

[69] Goyer R. Factors influencing metal toxicity. *Metal toxicology Academic Press, San Diego*. 1995:31-45.

[70] Brzóska MM, Moniuszko-Jakoniuk J. The influence of calcium content in diet on cumulation and toxicity of cadmium in the organism Review. *Archives of toxicology*. 1997;72(2):63-73.

[71] Florianezyk B. Toxic and cancerogenic properties of cadmium. *Nowiny Lekarskie*. 1995;64:737-45.

[72] Videla LA, Fernández V, Tapia G, Varela P. Oxidative stress-mediated hepatotoxicity of iron and copper: role of Kupffer cells. *Biometals*. 2003;16(1):103-11.

[73] Yiin S-J, Sheu J-Y, Lin T-H. Lipid peroxidation in rat adrenal glands after administration cadmium and role of essential metals. *Journal of Toxicology and Environmental Health Part A*. 2000;62(1):47-56.

[74] Müller L. Consequences of cadmium toxicity in rat hepatocytes: mitochondrial dysfunction and lipid peroxidation. *Toxicology*. 1986;40(3):285-95.

[75] Gaetke LM, Chow CK. Copper toxicity, oxidative stress, and antioxidant nutrients. *Toxicology*. 2003;189(1):147-63.

[76] Sgherri C, Quartacci MF, Izzo R, Navari-Izzo F. Relation between lipoic acid and cell redox status in wheat grown in excess copper. *Plant Physiology and Biochemistry*. 2002;40(6):591-7.

[77] Nikolić RS, Krstić NS, Nikolić GM, Kocić GM, Cakić MD, Anđelković DH. Molecular mechanisms of beneficial effects of lipoic acid in copper intoxicated rats assessment by FTIR and ESI-MS. *Polyhedron.* 2014;80:223-7.

[78] Suzuki YJ, Tsuchiya M, Packer L. Thioctic acid and dihydrolipoic acid are novel antioxidants which interact with reactive oxygen species. *Free Radical Research.* 1991;15(5):255-63.

[79] Arivazhagan P, Thilakavathy T, Ramanathan K, Kumaran S, Panneerselvam C. Effect of DL-α-lipoic acid on the status of lipid peroxidation and protein oxidation in various brain regions of aged rats. *The Journal of nutritional biochemistry.* 2002;13(10):619-24.

[80] Matsugo S, Yan L-J, Konishi T, Youn H-D, Lodge JK, Ulrich H, et al. The lipoic acid analogue 1, 2-diselenolane-3-pentanoic acid protects human low density lipoprotein against oxidative modification mediated by copper ion. *Biochemical and biophysical research communications.* 1997;240(3):819-24.

[81] Jovanovic JM, Nikolic RS, Krstic NS, Kocic GM. Monitoring of Lipoic Acid Protective Role by Liver Endonucleases Activity in Acute Intoxicity with Cadmium and Lead. *European Journal of Pharmaceutical Sciences.* 2011;44:186-7.

[82] Quinlan GJ, Halliwell B, Moorhouse CP, Gutteridge JM. Action of lead (II) and aluminium (III) ions on iron-stimulated lipid peroxidation in liposomes, erythrocytes and rat liver microsomal fractions. *Biochimica et Biophysica Acta (BBA)-Lipids and Lipid Metabolism.* 1988;962(2):196-200.

[83] Tandon S, Singh S, Prasad S, Srivastava S, Siddiqui M. Reversal of lead-induced oxidative stress by chelating agent, antioxidant, or their combination in the rat. *Environmental research.* 2002;90(1):61-6.

[84] Busse E, Zimmer G, Schopohl B, Kornhuber B. Influence of alpha-lipoic acid on intracellular glutathione in vitro and in vivo. *Arzneimittelforschung.* 1992;42(6):829-31.

[85] Gurer H, Ozgunes H, Oztezcan S, Ercal N. Antioxidant role of α-lipoic acid in lead toxicity. *Free radical biology and medicine.* 1999;27(1):75-81.

[86] Iakovidis I, Delimaris I, Piperakis SM. Copper and its complexes in medicine: a biochemical approach. *Molecular biology international.* 2011;2011.

[87] Ookhtens M, Kaplowitz N, editors. Role of the liver in interorgan homeostasis of glutathione and cyst (e) ine. *Seminars in liver disease;* 1997.

[88] McIntyre T, Curthoys N. Renal catabolism of glutathione. Characterization of a particulate rat renal dipeptidase that catalyzes the hydrolysis of cysteinylglycine. *Journal of Biological Chemistry.* 1982;257(20):11915-21.

[89] Orrenius S, Ormstad K, Thor H, Jewell S, editors. Turnover and functions of glutathione studied with isolated hepatic and renal cells. *Federation proceedings;* 1983.

# INDEX

## C



## D

## E





## F



## G



## H

## I





## K



## L

## O





## P

survival, 24
swelling, 44
symptoms, 15, 66
synaptic plasticity, 42, 45, 52, 53
synaptic strength, 42
synaptic transmission, 42, 53
syndrome, 36, 37, 38, 47, 48, 50
synthesis, 39, 62, 63, 73, 78

## T

target, 34, 45
teeth, 65, 79
temporal lobe, 15, 21, 26
Thailand, 33, 45
therapeutic use, 64
therapeutics, 78
therapy, 68, 78
threonine, 43
tissue, 2, 9, 10, 12, 14, 20, 47, 62, 65, 67, 68, 69, 70, 71, 72, 74, 75, 77
toxic effect, viii, 55, 56, 58, 65, 66, 71, 72, 74, 79
toxic metals, 59
toxic substances, 63
toxicity, 9, 16, 17, 19, 22, 24, 56, 65, 66, 68, 69, 75, 79, 80, 81
toxicology, 79, 80
transition metal, 2, 75, 78, 79
transition metal ions, 79
transmission, 42
treatment, viii, 9, 14, 15, 24, 26, 46, 51, 56, 80
triggers, 2, 47, 56
triiodothyronine, 49
tumor, 75
tumor cells, 75
turnover, 76
type 2 diabetes, 45, 46

## U

ulcerative colitis, 77
underlying mechanisms, 44
uric acid, 12, 61

## V

valence, 58
varieties, 11
vasculature, 6
VEGF expression, 49
vitamin C, 12, 22, 61
vitamin D, 66
vitamin E, 61
vitamins, 12, 68

## W

water, 41, 57, 65
weight control, 36
weight gain, 34
weight loss, 34, 45
Wistar rats, vii, viii, 51, 55, 56, 68, 70, 74, 79
Wnt signaling, 38, 50
workplace, 65

## Y

yield, 3, 5

## Z

zinc, 78, 80

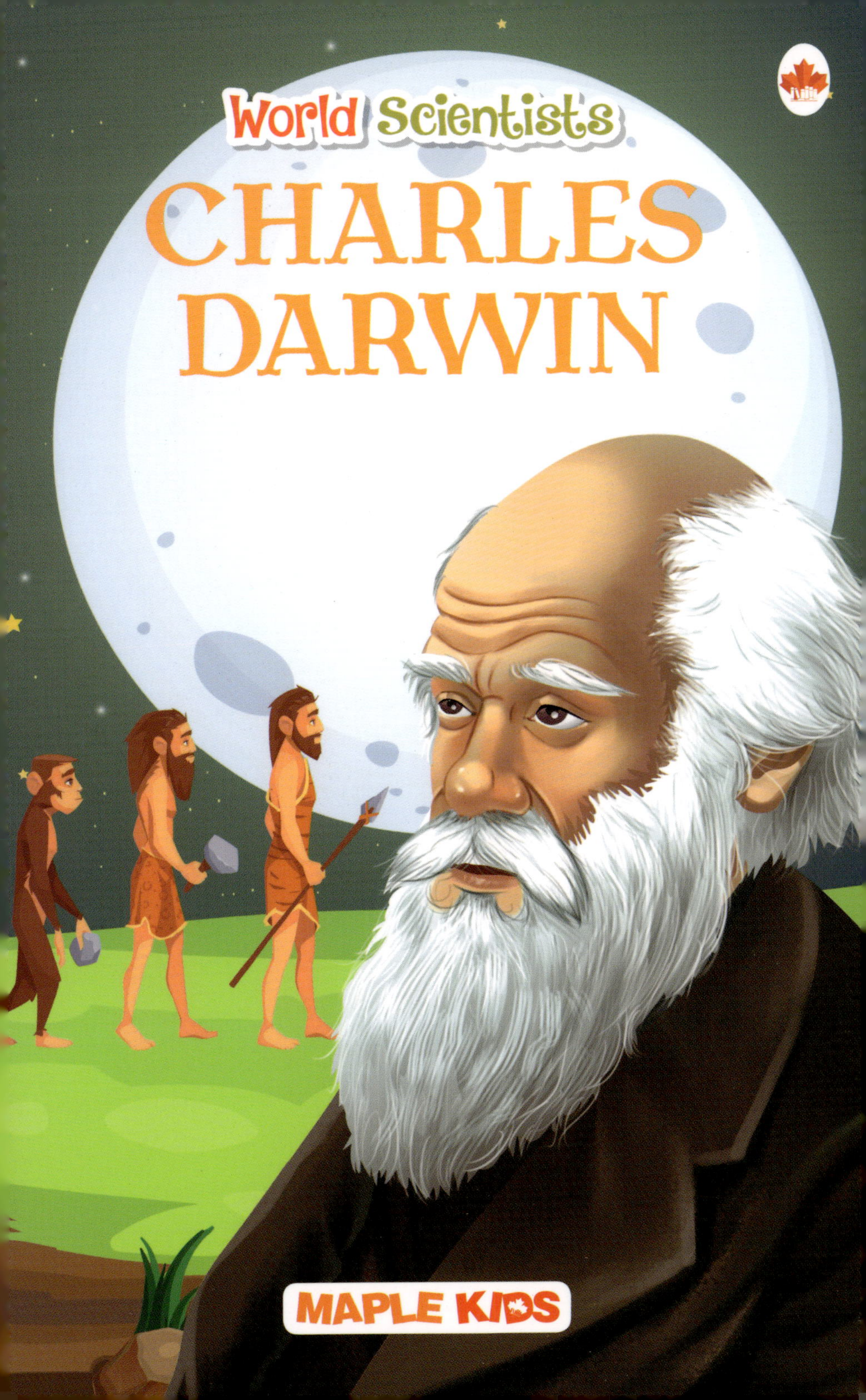

# World Scientists

# CHARLES DARWIN

Maple Kids